Trelawney Saunders

An Introduction to the Survey of Western Palestine

Its Waterways, Plains & Highlands

Trelawney Saunders

An Introduction to the Survey of Western Palestine
Its Waterways, Plains & Highlands

ISBN/EAN: 9783337140953

Printed in Europe, USA, Canada, Australia, Japan

Cover: Foto ©Andreas Hilbeck / pixelio.de

More available books at **www.hansebooks.com**

AN INTRODUCTION

TO THE SURVEY OF

WESTERN PALESTINE:

Its Waterways, Plains, & Highlands.

BY

TRELAWNEY SAUNDERS,

GEOGRAPHER.

ACCORDING TO THE SURVEY

CONDUCTED BY

LIEUTENANTS CONDER & KITCHENER, R.E.

FOR THE PALESTINE EXPLORATION FUND.

London:
RICHARD BENTLEY AND SON, NEW BURLINGTON STREET,
Publishers in Ordinary to Her Majesty.
1881.

CONTENTS.

PREFACE.
Limits of the Survey; Area; and Time devoted to it. The results. The Large Map. The Reduced Map. Special Plans of Towns, etc. Memoirs. General Index. Three editions of the Reduced Map. Character of the Survey. Its extension urged. The New Survey and former Maps.

THE SURVEY OF WESTERN PALESTINE. Introduction. The OUTLINE of the Map: Waterways, Basins, Waterpartings, and Watersheds. The Survey included in two Great Watersheds, that of the Mediterranean Sea on the West, and that of the Jordan on the East. The OROGRAPHY: Lowlands and Highlands.

PART I.

THE MEDITERRANEAN WATERSHED.   PAGE

THE BASIN OF NAHR KASIMIYEH. *Outfall*, 33° 20' 16" *N. Lat.* ...   9-12
The Southern Waterparting. Wady Hajeir. Wady Selukieh. Wady 'Aizakaneh. Rectifications of former Maps. Two classes of Basins. Out of 30 Basins along the Coast, only four drain the Upper Plateau of the Interior.

THE BASIN of WADY EL HUBEISHIYEH. *Outfall*, 33° 17' *N. Lat.*   12-14
Waterparting. Former Names and Misrepresentations. Two Main Channels: the Hubeishîyeh draining the centre and south, and the Humranîyeh draining the northern part. Wady Ashûr. Wady el Ma. Castle of Tibnin or Toron.

THE BASIN OF WADY EL EZZIYEH, AND MINOR BASINS ON THE NORTH. *Outfall of Wady el Ezzîyeh*, 33° 11' 40" *N. Lat.*...   14-16
Position. Previous inaccuracy of the Minor Basins. Wady el 'Akkâb. Wady Nettarah. Contracted Limits of the Lower Part. Expansion of the Upper part. Waterparting. Jebel Adâther. Khurbet Belât. Tell Belât. Rectifications of former Works.

A 2

iv   CONTENTS.
                                                                PAGE
THE BASIN OF WADY KERKERA. *Outfall*, 33° 4' 48'' *N. Lat.*   .. 16–18
Overlooked formerly.

THE BASIN OF WADY EL KURN. *Outfall*, 33° 3' *N. Lat.*   18–19
Relation to Wady el Ezzîyeh. Waterparting. Contact with Jordan
  Basin. The lower Gorge. Kulat el Kurein or Castle of
  Montfort. Two main branches. Jebelet el Arûs. Jebel
  Jurmuk. El Bukeiah. Gorge of Suhmata.

THE BASIN OF NAHR N'AMEIN AND MINOR BASINS ON THE NORTH   20–23
Outfall, 32° 54' 30'' *N. Lat. Minor Basins* between N'amein and
  Kûrn. Semeirîyeh, Majnuneh. Castle Jiddin. Nahr Mefshukh.
  Villages. Fountain el Kabry. Wady es Salik. *Nahr N'amein
  Basin.* Waterparting. Southern Range of Upper Galilee.
  Extent of Basin. Wady el Halzûn and Wady Shaib. Wady esh
  Shâghûr. Wady el Waziyeh. Base of Southern Range of
  Upper Galilee, and the highway between Acre and the East.
  Wady 'Abellîn. Three natural divisions of Lower Galilee.
  Rectifications.

THE BASIN OF NAHR EL MUKUTT'A (KISHON). *Outfall*, 32° 49'
  N. Lat... 23–26
The waterparting. Outfall. Wady el Melek, and the Northern
  Divisions of the Basin. Four Main Channels of the Southern
  Division. Affluents of the Central Division.

MOUNT CARMEL 26–30
Its eastern and western slopes, paths, wadys, main ridge, and
  parallel ranges.

THE BASINS OF NAHR ED DUFLEH (¹), AND NAHR EZ ZERKA (²)
  Outfalls (¹) 32° 35' 42''—(²) 32° 32' 20'' *N. Lat.*.. 30–32
(1.) Three permanent Streams. El Khashm. Plain of Tanturah.
  Depression of Belad er Ruhah, dividing Mount Carmel from
  Jebel Sk. Iskander, etc. (2.) Two divisions of the Nahr ez
  Zerka Basin. Notable rectifications.

THE BASIN OF NAHR EL MEFJIR. *Outfall*, 32° 28' *N. Lat.*   .. 32–35
Formerly Nahr Akhdar, and misunderstood. Merj el Ghûrûk.
  Wady es Selhab, Plains of Dothan and 'Arrabeh, Wady el
  Ghamik and Wady Abu Nar. The Nuzlet Villages. Wady
  el Maleh, Wady el Khudeirah, and Nahr el Mefjir. Wady er
  Roz, and Wady Abu Kaslan, Sheikh Beiazid Mountains.
  Wady Arah. Wady el Asl, and Wady Bir Isîr. Wady
  Samantar, Wady el Yahmur, and Wady Rasein. The Upland
  of Umm el Khatâf. Rectifications.

THE BASINS OF NAHR ISKANDERUNEH, AND NAHR EL FALIK. *Out-
  falls*, 32° 23' 50'' *and* 32° 16' 30'' *N. Lat*...   35–37
Outfall of Nahr Iskanderuneh and Waterparting. Rectification.
  Nahr el Falik. Wady esh Shair. Nablûs. Mount Ebal or
  Eslamiyeh. Wady Zeimer. Jebel et Tûr or Mount Gerizim.
  Wady et Tin. Wady el Haj Musa. Wady Kulunsaweh.
  Wady en Naml. Highways.

CONTENTS. v

PAGE

THE BASIN OF NAHR EL 'AUJA. *Outfall, 32° 6' 10" N. Lat.* .. 37–45
Outfall. Permanent streams. Waterparting. Merj Sia. Northern and Southern Divisions, and their partition. *Northern Division* —Wady Kalkilieh. Wady Azzun. Rectification. Wady Kanah, the biblical Brook Kanah. Wady Rabah. Rectification. Wady Ballût, Wady Bushanit and Wady Ishar, Wady er Rumt, Wady el Kub, Wady Seilun, Wady en Nimr, Wady el Jib, Wady er Reiya. Wady Suhary. *Southern Division*— Wady Nusrah. Khallet es Salib. Wady Shâhin. Wady esh Shellal or Budrus, and Wady Ludd, their partition. Tributaries of Wady esh Shellal :—Wady en Natuf, Wady Malâkeh, Wady Hamis, Wady Kelb, Wady Delbeh, and Wady Shamy. Wady Ain 'Arik, Wady el Imeish, Wady es Sunt, Bethhoron. Wady Muslih. Tributaries of Wady Ludd :—Wady Harir. Wady 'Atallah. Wady Aly. Wady Jaar. Wady Suweikeh. Wady Selman. Wady el Mikteleh. Wady el Burj. Wady Mozarki. Wady el Hai. Wady Khushkush. Wady Aly. Wady Alakah. Rectifications.

THE BASIN OF NAHR RUBIN. *Outfall, 31° 56' 10" N. Lat.* 45–50
Outfall. Waterparting. Curvature of the Basin. Headwaters on the Plateau of el Jib, or Gibeon. Wady ed Deir. Wady 'Amir. Wady Jillan, ed Dumm, and Beit Hannina. Wady el Abeideh. Wady Buwai. Rectifications around Jerusalem. Wady es Surar. Wady esh Shemarin. Wady es Sikkeh, Wady el Werd, and Wady Bittir. Wady Ismain. Wady en Nagil. Wady el Ghurab. Mount Seir, and Mount Ephron. Separation between the Mountains of Judah and the lowland Hills of of the Shephelah. Samson's Country. Plain of Akir (Ekron). Valley of Sorek. Wady Deiran. Wady Ayun Do. Wady en Nahir. Wady el Menâkh. Mount Baalah.

THE BASIN of NAHR SUKEREIR. *Outfall, 31° 49' 10" N. Lat.* .. 51–54
Waterparting. Three main branches. 1. Wady es Sunt. Wady el Jindy. Wady es Sur. Division of the Mountains and Lowland. Cave of Adullam. Gorge of es Sunt. Tell es Safi. 2. Wady el Afranj. Wady Kaideh. Rectification. 3. Wady ed Dawaimeh. Wady el Ghueit. Outfall. Rectifications.

THE BASIN OF WADY EL HESY. *Outfall, 31° 36' 20" N. Lat.* .. 54–56
Outfall. Waterparting. Basin of Wady Bireh. Main Sources and Channel of Wady el Hesy. Wady edh Dhikah, and en Nûs. Gheith, the site of Gath. Huj.

THE BASIN OF WADY GHUZZEH (GAZA). *Outfall, 31° 27' 54" N. Lat.* 56–61
Waterparting. Wady el Khulil. Wady el 'Aawir. Yutta. Wady Kilkis. Wady ed Dilbeh. Wady Deir el Loz. Wady Itmy. Wady es Seba. Ziph and Carmel. Wady el Khan. Wady el Habûr. Wady el Ghurra. Wady Saweh. Wady es Seba. Wady el Khureitein. Wady es Seba. Wady esh Sherîah. Shaaraim. Hazor Susah. Darum and Bizjoth-Jah-Baalah.

PART II.

THE WESTERN WATERSHED OF THE JORDAN AND DEAD SEA BASIN.

PAGE

THE BASIN OF WADY ET TEIM AND NAHR HASBANY. *Oufall into Huleh Marsh at es Salihiyeh,* 33° 10′ *N. Lat.* 62–64
The head of the Jordan Basin. Wady et Teim and Nahr el Hasbany. Merj 'Ayun. Perennial Fountains of the Jordan. Tell el Kâdy. Dan. 'Ain and Nahr Leddan. Banias, fount and stream. Nahr Bareighit.

THE HULEH PLAIN, MARSH, AND LAKE. *Outfall,* 32° 2′ 16″ *N. Lat.* 64–65
The affluents of the Huleh from Kades and Meis. Wady Arûs. 'Ain el Mellahah.

THE BASIN OF WADY EL HINDAJ. *Outfall,* 33° 3′ 3″ *N. Lat.* .. 65
The Waterparting.

THE MINOR BASINS OF WADY SHEBABIK (WAKKAS), ETC. 65–66
Wady Musheirifeh. Wady Loziyeh. Wady Zuhtuk. Wady Jamus.

THE BASIN OF WADY 'AMUD, SAFED. *Outfall,* 32° 51′ *N. Lat.* .. 66

THE BASIN OF WADY RUBUDIYEH. *Outfall,* 32° 50′ 30″ *N. Lat.* .. 66

THE BASIN OF WADY EL HAMAM. *Outfall,* 32° 49′ 50 *N. Lat.* .. 66–67

THE BASIN OF WADY FEJJAS. *Outfall,* 32° 41′ 50″ *N. Lat...* :. 67

THE BASIN OF WADY EL BIREH. *Outfall,* 32° 36′ *N. Lat.* 67-68
Wady Dabu, Yebla, or Esh-sheh.

THE BASIN OF NAHR JALUD. *Outfall,* 32° 31′ *N. Lat.* 68–70
'Ain Jalûd. Zerin (Jezreel). Beisan (Bethshean). Head of the Basin. Valley of Jezreel. Wady el Hufiyir. Wady es Sidr. 'Ain Tub'aûn or Tubania. Kunat es Sokny. Wady el Hariyeh.

THE BASIN OF WADY SHUBASH. *Outfall,* 32° 24′ 30″ *N. Lat.* . .. 70
Elevated glens, or lateral valleys of el Mughair and Raba. Ka'aun, the ancient Coabis.

THE BASIN OF WADY EL KHASHNEH. *Outfall,* 32° 23′ 35″ *N. Lat.* 70–71

THE BASIN OF WADY EL MALEH. *Outfall,* 32° 21′ 25″ *N. Lat.* .. 71-73
Waterparting. Three main branches:—Wady el Maleh and Castle. Wady Helweh. Wady ed Duba.

MINOR BASINS BETWEEN WADY EL MALEH, THE JORDAN, AND WADY FAR'AH 73–76
The Slope. Wady Umm el Kharrubeh. Wady Shaib. Wady el Bukei'a and Abu Sidreh. The Ghor and the Zor. Sh'ab el Ghoraniyeh.

CONTENTS. vi

	PAGE

THE BASIN OF WADY FAR'AH. *Outfall, 32° 2′ 20″ N. Lat.* .. 76–81
Waterparting. Dimensions. Head of the Basin in two parts. 1.
Southern part—Plains of Rûjib, Askar, and Salim, Chasm of
Wady Beidan. 2. Northern part—Two centres at 'Ain and
Tell Fâr'ah, and Wady Beidan. Wady Fâr'ah. Sea Level.
Fall. Three divisions. El Fersh. Buseiliyeh. Gorge. The
Kurawa. Masudy Arabs. Archelaus. Affluents. Shab esh
Shinar. Nukb el Arais. Wady el Jozeleh. Route of Abram,
Jacob, and Benhadad the Syrian.

THE BASIN OF WADY EL HUMR. *Outfall, 32° 1′ 30″ N. Lat.* .. 81–84
Waterparting or Boundary. Wady el Humr. Wady el Kerad.
Wady Zamur and ed Dowa. Gorge. Wady el Ifjim. Precipitous Chasm. Wady Zakaska. Kurn Surtubeh. Wady Fusail.
Phasaëlis.

THE BASIN OF WADY EL 'AUJAH. *Outfall, 31° 55′ 10″ N. Lat.* .. 84–87
Waterparting. Three divisions. (1.) Wady el Mellahah. Wady
Unkur edh Dbib. Wady Bakr. Wady Mekûr edh Dhib.
Change in the base of the Mountains. Expansion of Low Hills.
Intermediate Plain, probably the Plain of Keziz. (2.) Wady
el 'Aûjah. 'Ain Samieh. 'Ain el 'Aûjah. Wady Abu el
Haiyat. Wady Sebata. Wady el Abeid. Wady en Nejmeh.
Wady Dar el Jerir. Wady Lueit. (3.) Wady Abu Obeideh.
Wady Umm Sirah. Rectifications.

THE MINOR BASIN OF WADY MESA'ADET 'AISA. *Outfall, 31° 53′ 10″ N. Lat.*... 87
A Traditional Mountain of the Temptation.

THE BASIN OF WADY NUEI'AMEH. *Outfall, 31° 52′ 28″ N. Lat.* ... 88–90
Sources. Waterparting or Boundary. Curvature of the Basin,
and its effect on Lateral Communication. Watercourses. Wady
el 'Ain. Khallet es Sultan. Wady el Kanabis. Wady Muheisin. Wady Asis. Rummôn ("the Rock Rimmon"). Wady
Abu el Haiyat, and Wady el Asa. Wady es Sineisileh. Wady
Rummâmaneh. Wady el Harik. Wady el Makûk. Wady
Abu Jurnan. 'Ain en Nûei'ameh. 'Ain ed Dûk. Osh el
Ghurab. El Ghoraniyeh Ford.

THE BASIN OF WADY EL KELT. *Outfall, 31° 49′ 30″ N. Lat.* .. 90–97
Waterparting. Extent. Watercourses in two parts. (1) Wady
Suweinit. Ai. Wady el Medineh. Wady en Netîf. Jeba (Geba).
Mukhmas (Michmash). Site of Philistine camp. The Rocks
Bozez and Seneh. (2) Wady Fârah. Wady Redeideh. Wady
esSenam. Wady Zimrij. Wady en Nimr. Wady en Nukheileh.
'Ain el Kelt. Wady el Kelt. Wady Abu Duba. Khan Hathrurah
and Talat et Dumm on the Jericho Road. Defile. Jericho. Khaur
Abu Dhahy. 'Ain Hajlah. Wady Rijan. Roads and tracks.
Valley of Zeboim or the Hyenas' Ravine identified with Wady
Sikya. Note on Ai.

THE BASIN OF EL KUEISERAH AND MINOR BASINS ON THE
NORTH. *Outfall of el Kueiserah, 31° 45′ 40″ N. Lat.* 97–102
Formerly Wady Dabor. Slight contact with Mediterranean Slope.

CONTENTS.

PAGE

Waterparting or boundary. Minor Basins falling into the Dead Sea. Wady Talat et Dumm. Wady Mukarfet Kattum. Wady Joreif Ghuzûl. Watercourses of el Kueiserah Basin. Wady Seleim. Gorges of Deir es Sidd. Wady Ruabeh from Mount of Olives. Wady es Sidr. Pass of Thogret ed Debr. Wady el Lehhâm from Mount of Olives. Wady el Haud. Wady el Jemel. Jebel Ekteif. Wady el Mudowerah, and Wady ed Dedakin. Roads. Gorge of el Mukelik. David's Flight from Absolom.

THE BASIN OF WADY EN NAR (BROOK KIDRON) AND THE MINOR BASINS ON THE NORTH. *Outfall of Wady en Nar*, 31° 40′ 20″ *N. Lat.* 103–108

Wady Jofet Zeben. Wady Kumran. Plateau of el Bukei'a. War ez Zeranik. 'Ain' Feshkah. Wady es Sammârah. Waterparting of Wady en Nar Basin. Valley of Jehoshaphat and the Valley of Hinnom. Bir Eyûb. Wady Abu Aly. Mar Saba. Roads from Jerusalem to Mar Saba. Wady Akhsheikh. Wady Jerfân. Wady 'Alya. Wady el Areis. Wady Umm eth Theleithât. Rectifications.

THE BASIN OF WADY ED DERAJEH *Outfall*, 31° 34′ 35″ *N. Lat.* 108–110

Wady el Ghuweir. Waterparting. Divisions. (1) Wady el Kaah. Wady Samurah. Wady Lozeh. Wady Umm el Kulah. Wady et Tin. Wady el War. Wady D'abûd and Wady el T'âmireh. Wady el Meshash. Wady el Bussah. (2) Wady el Biar. Valley of Berachah. Aqueduct. Wady Fureidis. Cliffs and Caves of Khureitun, the traditional Adullam. Wady Jubb Iblan. Chasm of Wady Muallak. (3) Wady Mukta el Juss.

THE BASIN OF WADY EL 'AREIJEH AND THE MINOR BASINS ON THE NORTH. *Outfall of Wady el 'Areijeh*, 31° 27′ 30″ *N. Lat.* 110–116

Outfall. Minor Basins :—Wady Hûsâsah, Wady esh Shukf, Wady Sideir, Wady Marjari. Roads from 'Ain Jidy. Waterparting or boundary of el 'Areijeh Basin. Watercourses of el 'Areijeh. Plateau of el 'Arrûb. Wady el "Arrûb, and its affluents. Gorge of Wady el Jihar and Wady el Ghar. Plateau of Jihar and Ghar. Wady Umm el 'Ausej. Wady es Suweidiyeh and Wady es Sûkiyeh. Route.

THE BASIN OF WADY EL KHUBERA. *Outfall*, 31° 24′ *N. Lat.* 116–117

Waterparting or boundary. Three divisions. (1) Wady Jerfân. Wady el Kuryeh. Wady Nimr. Wady es Sihaniyeh. Wady Umm Kheiyirah. (2) Wady Malâki. Wady Kueiwis. Wady el War. Malâki and Khubera Gorges. (3) Wady Rujm el Khulil. A scene of David's exploits.

THE BASIN OF WADY SEIYAL AND MINOR BASINS ON THE NORTH. *Outfall of Wady Seiyal*, 31° 19′ 47″ *N. Lat.* 118–120

Wady Mahras. Wady el Kasheibeh. W. Sufeisif. Waterparting of Wady Seiyal Basin. Wady Khurbet et Teibeh. Wady el Kurcitein. Wady es Sennein. Wady Mutan Munjid. Wady Umm Jemat. Wady el Khuseibiyeh.

PART III.

THE PLAINS OF WESTERN PALESTINE.

1. THE PLAINS ON THE MEDITERRANEAN SLOPE.

PAGE

THE PLAINS OF GALILEE 121-131

The Maritime Plain of Tyre. Ras el Abyad to Ras en Nakura. The Maritime Plain of Acre and the Plain of Megiddo. Intermediate Range of Hills. The Plain of Rameh. The Plain of 'Arrâbeh. The Plain of Buttauf. The Plain of Tôran. General view of the Plains of Acre and Megiddo. Six great arms. (1) The Plain of Acre. (2) W. Halzun, and the plains of Rameh and 'Arrâbeh. (3) The plains of Buttauf, Toran, and Seffurieh. (4) Between Mt. Tabor and Jebel Duhy. (5) Nahr Jalûd, or the Valley of Jezreel. (6) Between Mt. Gilboa and Mt. Ephraim. The Central Plain. Roads.

THE MARITIME PLAINS, SOUTH OF CARMEL 131-142

(1) The Plain of Tantura. (2) The Plain of Sharon. Forest of Oaks. Falik Hills. Dhahr Selmeh Hills. Ramleh Hills. Three divisions of the plain of Sharon. (3) The Plains of 'Arrâbeh or Dothan. (4) The Plains of Mukhnah, Rujib, 'Askar, and Salim. (5) The Plain of Philistia. Line of highland prolonged. Plains at Akir and 'Arâk. Intermediate hills and hollow. Plain of Akir or Ekron, and plains west of Jerusalem. Plain of Yebnah (Jabneel). Sandy downs. Plain of 'Arâk el Menshiyeh, and Heights of Hebron. Ashdod. Advance of Hills towards the coast. The Sukereir Range. The Keratiya or Bureir Valley. Askelon. Site of the Port of Askelon. The guardianship of the Five Cities, and the site of Gath.

2. THE PLAINS, LOWLANDS, AND LAKES OF THE JORDAN SLOPE.

THE PLAIN OF MERJ 'AYUN 143-144

Ijon. Ain Derderah. Nahr Bareighit. Deadly Storm. Altitude. Relation to the Kasmîyeh or Litany River and the River Jordan.

THE HULEH PLAIN, MARSH, AND LAKE 144-147

Total dimensions. Four divisions. (1) Huleh Plain. Surrounding heights. (2) Huleh Marsh. Dr. Tristram's notes. Mr. J. Macgregor's boat survey. Channels. Dangers. (3) Huleh Lake. Form and extent. Altitude. Soundings. (4) Ard el Kheit.

THE GALILEAN GORGE 147-148

Fall of the Jordan. Bridge of Jacob's Daughters. Western affluents. Eastern end of the Southern Range of Lower Galilee. Its relation to the depression of the Jordan.

CONTENTS.

	PAGE
THE WESTERN SHORE OF THE SEA OF GALILEE	148-152

Entrance of the Jordan. Plain of Batihah. Bethsaida Julias. Tell Hum. Kh. Kerayeh. Et Tabghah, Bethsaida of Galilee. Plain of Gennesaret. Interesting Wadys. 'Ain 'el Mudauwerah, Capernaum placed at Khan Minia. Magdala at el Mejdel. Chinnereth at Abu Shusheh. Wady Abu el 'Amis and the Ayun el Fuliyeh. Merj Hattin. Wady el Hamâm. Irbid, Arbela, or Beth Arbel. Kulat Ibu Mân, fortified Caverns. Hajaret en Nusara. Horns of Hattin. Tiberias. Emmaus. Lofty Cliffs. Kerak. Exit of the Jordan.

THE GHOR FROM KERAK TO JISR MUJAMIA	152-153
JISR MUJAMIA TO NAHR JALUD	153-154

Wady Birch. Wady el Esh-sheh. The Zor.

THE PLAIN OF BEISAN	155-159

Nahr Jâlûd and the Valley of Jezreel. Mt. Gilboa. Remarkable bank. Altitudes. Irrigation Channels. Wady Shubash. Wady el Khashneh. Mounds. Succoth. Lieutenant Conder's Megiddo.

THE SAMARITAN GORGE OF THE JORDAN	159-161

Relation to Wady Maleh. Contraction of the Ghor and Zor. Ten fords. Comparative lists of names.

THE PLAIN OF PHASAELIS..	162-166

Northern limit. Three parallel valleys connected with the plain. (1) Wady Sidreh. Wady el Bukeia and its peculiarities. El Makhrûk. Archelaus. (2) The Kurawa and Wady Fâr'ah. Kurn Surtabeh. Recess. End of long line of cliffs. (3) Wady Ifjim and its peculiarities. Khurbet Fusail, the site of Phasaëlis. Wady el Humr. Wady el 'Aûjah. Expansion of low hills across the plain. Recess of the Mountains and enclosed plain. 'Ain Dûk. 'Ain en Nûei'ameh. Vale or Plain of Keziz. Special features. Wady el Jozeleh. Wady el Mellâhah. Mankattat or "Strips." Wady Mesâ'addet 'Aîsa or the "Ascension of Jesus." 'Osh el Ghurâb or the Raven's nest. Traditional sites of the Mountain of Temptation. Zemaraim.

THE PLAIN OF JERICHO	166-170

Limits. Kusr el Yehûd or Jew's Castle. The Zor. Mouth of the Jordan. Dead Sea coast. Mountain and Cliffs. Jebel Kuruntul. Erîha. Sites of ancient Jericho. 'Ain es Sultan. Wady Kelt. Interruption of the line of Cliffs. Slope of plain. Fertility. Wadys. Wady Nûei'ameh. Wady Kelt. Jiljulieh, the site of Gilgal. Ain Hajlah, site of Beth Hoglah. Kusr Hajlah. Kusr el Yehûd or Jew's Castle, a ruined convent. The Pilgrims' Bathing places. Reputed sites of Our Lord's Baptism. Greeks and Latins. Khaur el Thumrâr. Wady Makarfet Kattûm. Wady Joreif Ghuzal. Wady el Kaneiterah. Wady Kumran. Gomorrah.

CONTENTS.   xi

3. THE WESTERN SHORE OF THE DEAD SEA.

                    PAGE

RAS FESHKAH TO RAS MERSID 170–172

Dr. Tristram's Map and the New Survey. Ras Feshkah. The Shore. Sulphur Spring. The intersecting Wadys. Plain of Engedi. Wady Sideir. Wady el Areijeh. Pass, plateaus, spring, and vast cliffs of 'Ain Jidy (Engedi, Cliff of Zor).

RAS MERSID TO SEBBEH 172–174

The Shore. Terraces. The Rock of Sebbeh. Ruins of Maasad. Wady Kuberah. Wady Seiyal.

PART IV.

THE HIGHLANDS OF WESTERN PALESTINE.

INTRODUCTORY 175–177

Relations of Streams and Plains to the Highlands. The Culminating Summits. Ranges. Edges of Plateaus. Altitudes.

THE MOUNTAINS OF UPPER GALILEE.

THE SOUTHERN RANGE 177–178

Summits. Base. Choice between Wady el Waziyeh with Wady esh Shaghûr, and Wady el Halzûn with Wady Shaib. Wady el Tawahin or the Safed Gorge.

THE EASTERN RANGE 178–189

Summits. Altitudes wanting. Slope. Altitudes along the base line. Peculiar Descent of the Jordan into the Huleh Plain. The line of lowest depression between Mount Hermon and the Mediterranean Waterparting. The slopes around the Huleh Plain. Dr. Robinson's six terraces. The western slope of Nahr Bareighit. The hypsometrical connection between Jebel Hunin, Abl, and Banias. The summit north of Jebel Hunin and south of it around the Basins of Meis and Tufeh to Neby Muheibib and Deir el Ghabieh. Parallel Range on the west of Wady 'Aizakaneh, continuous with the range running south from Neby Muheibib. Plateau of Kades. Terraces of Malkiyeh and Belîdeh. Wady 'Arûs. Ard el Dawamin. Ard el Kheit. Wady Selukieh, Wady Hindaj, and Wady 'Amûd. Jebel Kan'an.

THE WESTERN SLOPE AND ITS UPPER PLATEAUS 189–190

The lower slope interrupts passage parallel to the coast-road by its valleys and ridges. Parallel passage restored by the upper plateaus of the Hubeishiyeh, Ezziyeh, and Kûrn.

xii   CONTENTS.
                                                                PAGE
THE WESTERN RANGE   190–195
   Kisra. Line of heights. Corresponds with Waterparting of W. el
   Kurn. Gorge of the Kurn. Kulat el Kurein, the Crusaders'
   Montfort. Khurbet Belat. Line of Heights. Gorge of the
   Ezziyeh. Crusaders' Castle of Tibnin and its curious situation.
   Three divisions of the western slope. The Central Division. The
   Northern Division, upper and lower parts. Southern Division.

THE NORTHERN RANGE   196

THE INTERIOR OF THE PLATEAU OF UPPER GALILEE..   ..   197–199

   The Jermuk Range. The Marûn Range. The Main valleys
   of the Upper Kûrn. El Bukeiah. The Plains between the Jermuk and Marûn Ranges, and between the Marûn and the Eastern
   and Northern Ranges. Common base of the Jurmuk and Marûn
   Ranges.

           THE MOUNTAINS OF LOWER GALILEE.

INTRODUCTORY   199
   The Limits. Different aspects and altitudes of Upper and Lower
   Galilee.

THE NORTHERN OR SHAGHÛR RANGE   199–201
   Its base lines. It culminates in a plateau between two ranges. Altitudes wanting.

THE TORAN RANGE..   201–202
   Its base lines. Curvature. Scarps and cliffs. Plain of Hattin.
   Plain of Ahma. Culminating points. Hajaret en Nusara or
   Nazarite Rocks, a traditional site of "the Feeding of the Five
   thousand."

THE NAZARETH RANGE   202–211
   Its base lines. The Summits. *The Western Division.* Oak
   Forest. *The Central Division.* Semûnieh. Yafa and Mujeidil.
   El Warakany. Jebel Kafsy, the traditional Mount of Precipita-
   tion. Mount Tabor. Wady el Mady. Woods and chalk downs
   in contrast. Jebel es Sih. El Meshhed. Seffûrieh. Wady
   Kefr Kenna. *The Eastern Division.* Wady Shubbabeh and
   Wady Fejjas. Basaltic Cliffs. Depression of Base line below sea
   level. Wady Mu'allakah. The Plateau of Sh'arâh, its streams
   and villages. Successive descents from Mount Tabor to the Sea of
   Galilee.

THE JEBEL DUHY RANGE..   210–214
   The group, its limits and dimensions. Three divisions. (1) Jebel
   Duhy. Nein. Endor. Tell el 'Ajjûl. Eastern base. Kh. Maluf.
   El 'Afuleh and Aphek. Solam and Shunem. Saul's last battle.
   Tub'aûn and Tubania. Camps of the Crusaders and Saladin.
   Kleber's decisive victory. Eastern Plateau diagonally divided by
   Wady Dabu and Yebla. (2) Northern range with the Kaukab el
   Hawa or Castle Belvoir. Bold features. (3) Wady Kharrâr.
   Four successive terraces from the northern range to Nahr
   Jâlûd.

CONTENTS. xiii

	PAGE
GENERAL VIEW OF LOWER GALILEE	214

Its basins, lowlands, ranges, uplands, plains, and scenery.

THE HIGHLANDS OF SAMARIA AND JUDÆA.

INTRODUCTORY 215–216

The Geographical analysis of these mountains impracticable before the present Survey. Dr. Robinson's aims and labours. The new Survey now supplies the materials. Five divisions adopted.

THE NORTHERN SAMARITAN HILLS 216–223

The North Eastern base. The Western face. The Eastern face. The northern and southern division distinctly marked. The eastern end of the dividing line indicated. The features on either side contrasted. The Wady el Ifjim. The eastern side of the northern part described. The prolongation of the dividing line described westward to the sea, from Wady el Ifjim, through the Plain of Salim, the Nablus gap, Wady esh Shair or Zeimer, and Nahr Iskanderuneh. The northern escarpment of the northern part, and its summit. *Mount Gilboa* or *Jebel Fukû'a*. Double Summits in the south. Three spurs at northern end. Broad western branch and lesser spur. Abrupt eastern slope. Elevated lateral valleys. Valley of Jelbon. The contortions of the waterparting. Valleys of el Mughair and Raba. Southern boundary indicated by the drainage. Summary view of the western division of the Northern Samaritan Hills in six sections. Connected plains. Routes.

THE SOUTHERN SAMARITAN HILLS 224–228

The Southern limit. *The Eastern Side* contrasting the Southern with the Northern Hills. Eastward projection of the waterparting. Merj Sia. Eastern side also contrasted with the next section on the south. Elevated lateral valleys. Jebir 'Akrabeh with upper and lower terraces. Wady Nasir. Wady ed Duba and Wady el Merâjem. *The Western Side.* Undulating plateau. Western and Eastern altitudes. Plain of Mukhnah. Wady Ishar. Interior ranges. Wady Kanah. Falik Hills.

THE MOUNTAINS OF JUDÆA. (1) NORTHERN GROUP .. 229–231

Northern limits. Mount Ephraim. Summit of Northern Slope, culminating points. Shebtin heights and valleys. Lateral communications. Southern limits. Five Parallel ranges. Isolated hills in the Plain of Sharon.

THE MOUNTAINS OF JUDÆA. (2) THE JERUSALEM GROUP 231–241

Separation of the Mountains and Lowland Hills, on the south of Wady Malâkeh. Wady el Muslib. Wady el Mikteleh. Valley of Ajalon. Interruption and recurrence of Meridional valleys at Yalo. Wady en Najil. Wady es Sur. Hills and Mountains distinguished by altitudes. Plains of el Jib and Rephaim, or the Western Plateau of Jerusalem. Southern limits of the group. The low hills of this group distinct from the Shephelah. (1) Western Slope from Jerusalem Plateau. (2) Southern or Middle Slope. (3) Eastern or Jordan Slope. The Plain of Bukeia. Line of Eastern Summits. Eastern Plateau of Jerusalem. Middle Range.

CONTENTS.

	PAGE
THE MOUNTAINS OF JUDÆA. (3.) THE HEBRON GROUP	241-249

Limits. Main Range. Western Slope. Eastern Slope. The Middle Range continued. Eastern Range continued. Triple terraces. Highest Plateau. Wady el Khulil or valley of Hebron.

THE SHEPHELAH OR PHILISTIA 243-256

In five groups:—1. From Wady es Surar to Beit Dejan. 2. Between Wady es Surar and Wady es Sunt. 3. Between Wady es Sunt and Wady el Afranj. 4. Between Wady el Afranj and Wady el Hesy. 5. South of Wady el Hesy.

PREFACE.

THE first instalment of the Survey of the Holy Land, which the Committees and Subscribers of the Palestine Exploration Fund have for so many years persevered in producing, is at length published, and together with the work at Jerusalem and minor results, it may well be offered as a justification of all the exertions, and outlay, which have been expended.

This portion of the Survey of Palestine is bounded by the Nahr el Kasimîyeh or Litany River on the North, and by the Wady es, Seba on the South; with as much of the Jordan and the Dead Sea on the east, and of the Mediterranean Coast on the west, as the northern and southern limits admit. Within this extent are—the southern part of Phœnicia including Tyre but omitting Sidon; nearly the whole of Galilee; all Samaria; and the greater part of Judæa;—indeed from Dan to Beersheba.

The whole of the surveyed area covers more than 6,000 square miles. The survey occupied seven years in the field, and more than two years in addition were spent in the preparation of the work for publication. The immediate results of this survey in particular include: 1. A large map on the scale of one mile to an inch, reproduced and published in 26 sheets, each of which measures 22 inches by 18, the whole when joined together extending to 13 feet by 7. 2. A reduction of the large map for general purposes, on the scale of about 2½ miles to an inch, in six sheets, the size when joined together being 5 by 3 feet. 3. Numerous special plans on large scales of towns, buildings, and ruins. 4. Memoirs composed from the field notes of the 'surveyors, and from abstracts of authentic works. These treat on the natural features and products of the country, its hills and valleys, springs, wells, cisterns, water-courses, and streams; its present divisions, towns, villages, ruins, and highways;—the identification of Biblical and other Historical sites;—the inhabitants, their languages, legends and traditions, manners and customs, and superstitions. 5. A general index to the native names of about 9,000 places, with their signification as far as possible. 6. Photographs, sketches, and other illustrations. The plans, memoirs, index, and illustrations, will be combined and published in quarto volumes, now in the press. Three editions of the reduced map are in hand, illustrating: 1. The Modern Geography. 2. The Old Testament, and 3. The New Testament.

The publication of this remarkable survey with its accompaniments, goes far to throw open a splendid field of critical research, in the most satisfactory way. Without such a map of the country as the Fund has produced, the student of the History of Palestine, sacred and profane, ancient, mediæval, and modern alike, had to grope about in the midst of uncertainty. Even the most gifted explorers on the ground, could only partially note and record the facts connected with their actual route, while they had no adequate means of bringing the neighbourhood beyond their reach on either side, within the scope of their research. A map constructed on the basis of route surveys must be unequal to the requirements of modern science. It is but an imperfect substitute for such a work as the present. Few unsurveyed countries had received more able or more abundant attention than Palestine in the form of route surveys; these indeed being combined in its case with more pretentious works, for which neither adequate time nor proper arrangements had however been provided.* To form a judgment on the comparative value of the new survey with former publications, the best of which are the latest, but by no means the largest, it is only necessary to examine them together, in any part, when the inexpressible superiority of the new work will at once be observable.

* *E.g.*, Jacotin's map in five sheets, 1798-1801. Zimmermann's map in sixteen sheets, 1850.

To one who laboured with Dr. George Grove to produce a map of the Holy Land from the materials in existence before the Palestine Exploration Fund Survey, it may be allowed to express a lively and grateful sense of the merits of the new work. Unlike all its predecessors it is an original survey of the ground as a whole, on the best scientific methods. It is derived entirely from actual observation, carried on throughout on the same accurate and exhaustive basis; and it conveys all the information that is usually desired in a topographical map. It is a sound foundation for every kind of research in Palestine connected with topography. Such is the work that has been designed and executed by a body of private individuals depending on voluntary subscriptions. It excites the hope that the same successful instrumentality will now be directed to the completion of the survey of the Holy Land, including Lebanon on the north, the Negeb and Desert of the Wanderings on the south, and the interesting regions east of the Jordan. In the first "Statement of Progress" issued by the Committee, it was said that "so long as a square mile in Palestine remains unsurveyed, so long as a mound of ruins in any part, especially in any part consecrated by the Biblical history, remains unexcavated, the call of scientific investigation, and we may add the grand curiosity of Christendom remains unsatisfied." The spirit thus manifested cannot fail to derive great encouragement from the successful results of the labours which it has accomplished,—encouragement to persevere in the good work of applying the evidence existing on the land, to the elucidation of the record in the book. It is gratifying to add that since these remarks were in type, the Trans Jordan Survey has been decided upon.

A comparison of the new survey with former maps, displays an immense accession of detail, in such a form as to make one feel familiar with the country, and able to follow up the most obscure tracks along which any of the Biblical or other Historical narratives may lead. The map will doubtless elicit a general overhauling of the records of the past relating to the Holy Land, both sacred and secular. Next to visiting the scene of great exploits is the delight of tracing them on a faithful map of adequate scale, and even on the ground itself the map is a needful expositor. A map like the present, has all the character of a new revelation, and the exercise of the critical faculty, is sure to be brought into play on Palestine to an unprecedented extent, with regard to every interesting site, whether it have been identified, or have hitherto escaped identification. The discussion of such questions opens up a lively prospect for the editor and readers of future " Quarterly Statements."

THE SURVEY OF WESTERN PALESTINE.

INTRODUCTION.

To appreciate the use and value of the New Survey, it is necessary to examine it in some detail, and to bring it into comparison with the information that existed previously. Passing over the methods by which a survey is accurately made, it is enough at present to observe that the foundation of a geographical map is its Outline, consisting chiefly of the delineation of its Waterways. A distinct acquaintance with this branch of the subject is a fundamental element of geographical knowledge.

To describe the waterways of a country intelligibly, it is needful to adopt a simple method, based on the following facts. In tracing a watercourse from its source to its final outfall, it is found that some outfalls dispose of the waters of very small areas with simple systems of watercourses; while other outfalls are the drains of very large areas, with a great complication of watercourses, not easily unravelled. The area drained through each distinct outfall or mouth, is called a Basin, whether it be small or large; and its boundary is a Waterparting. The surfaces descending from a Waterparting to a Watercourse are called Watersheds or Slopes. The term Watershed or Slope is equally applicable to the sides of the smallest valley or of the largest continent. In the latter case, the term includes all the basins into which the same continental slope is divided. In like manner a Waterparting may be a simple ridge or mere swelling of the ground between two of the smallest streams, as well as the natural division of the waters of a Hemisphere.

The New Survey is included in two great Watersheds, that of the Mediterranean Sea on the west, and that of the Jordan on the east. The main Waterparting between these sheds or slopes, runs through the map between the Jordan and the Sea coast, in a very zig-zag course, which it is difficult to follow on the map, without prominent distinction by colour or by detailed description. About three-fourths of the country west of the Jordan are on the Mediterranean slope. The Basins of the Mediterranean Watershed will be examined first, then those of the Jordan. Afterwards the Orography or Relief of the country, its Plains and Highlands, will undergo a systematic examination, bringing out the distinctive forms of the ground, taking further notice of the Valleys, and rendering the intricate combination of the mountains intelligible. This will be the limit of the present work; but other investigations will follow.

THE SURVEY OF WESTERN PALESTINE.

PART I.

THE MEDITERRANEAN WATERSHED.

THE BASIN OF NAHR KASIMÎYEH.

Only a part of the basin of the Kasimîyeh, or Litany, falls within the Survey, its northern limit being chiefly the lower course of that river. The southern waterparting of the basin commences on the sea coast, about midway between Tyre and the mouth of the river. It is not well-defined at first, but it comes within a mile of the river on the west of Bidias, and here divides the Kasimîyeh from the Hubeishîyeh basin. The waterparting continues eastward as far as Sarifa, where it is a mile and a half from the river, but it bulges southward on the way, to a distance of two miles.

From Sarifa, the waterparting passes south-eastward to Burj Alawei, then southward to el Yehûdiyeh, eastward to beyond Safed el Battikh, southward round Berashît, and westward through Beit Yahûn to Ras et Tireh; where the Kasimîyeh basin ceases to be in contact with the Hubeishîyeh, and becomes contiguous with the Ezzîyeh basin. The waterparting runs on to Marûn er Ras, having the villages of et Tireh and Bint Umm Jebeil on the west, and those of Ainitha and Marûn er Ras on the east.

The Kasimîyeh basin now runs with that of the Jordan, and proceeds from Marûn-er-Râs east to Deir el Ghabiyeh, then north towards Meis where it makes another bend to the east, and then north to Hunin and el Khurbeh, separating the Wady 'Aizakaneh and other small branches of the Kasimîyeh from the Nahr Bareighut or Derdera and other tributaries of the Jordan. Beyond the survey, the Kasimîyeh rises on the eastern side of the highest part of Lebanon, on the Dhor el Khodib or Jebel Akkar, about 70 miles north of the great bend where it falls into the Palestine Exploration Map. It drains the central and south-western parts of the valley of Hollow Syria, lying between Lebanon on the west, and Anti Lebanon with Mount Hermon on the east. In the south-eastern part of Hollow or Cœle Syria, the Kasimîyeh is divided from Hermon by Jebel et Dahar and the Wady et Teim, in the basin of the Jordan. The Kasimîyeh basin is here confined to the eastern slopes of Tomat Niha, and the river runs in a vast chasm at the foot of that mountain, the twin peaks of which form the culminating summits of Southern Lebanon. At the southern end of that range, the chasm opens up into a small plain, and the basin expands a little, so as to include the Jermuk River which descends from the western side of Tomat Niha, and joins the Kasimîyeh about two miles north of Kulat esh Shukif or Belfort Castle. Here the river falls within the Survey, which, however, only extends to its southern bank, where it bends westward to the sea. On the north bank the basin is confined to narrow limits, and gives off only a few short branches to the main stream.

Within the Survey the basin of the Kasimîyeh is chiefly within the Wady el Hajeir and its tributary Wady Selukieh. The latter rises at Marûn er Ras and joins the Hajeir in a deep gorge on the east of Burj Alawei. The Hajeir rises at el Jumeijmeh, and falls into the Kasimîyeh near the Bridge of K'ak'aiyeh. The direct distance between Marûn er Ras and the Kasimîyeh exceeds fourteen miles. The width of this part of the basin varies from three to eight miles.

On the east of Wady el Hajeir, the Kasimîyeh, in bending to the west, receives the Wady 'Aizakaneh, from a valley on the south. The Wady 'Aizakaneh is in the same line as the main stream, before it bends to the west, but the course of the 'Aisakaneh is directly opposite. It rises near Hunin, and

THE BASIN OF NAHR KASIMÎYEH. 11

has a length of about six miles. Further east another valley divides the 'Aizakaneh from the waterparting of the basin.

On the west of the Hajeir, the southern margin of the Kasimîyeh basin contains several villages; and some small tributaries cut their way to the main stream through its high and rocky banks. One of the streams runs parallel to the waterparting by two channels for some distance.

The basin of the Kasimîyeh has been corrected as follows: The Hajeir is found to rise near el-Jumeijmeh, instead of at Aitheran. It is the Selukieh (Seluky), a branch of Hajeir which carries the basin of the Kasimîyeh so far south as Mârûn-er-Râs. West of Wady Hajeir the basin is deprived of a considerable area reaching south in former maps beyond Ter Zibna (now Teir Zinbeh), the new survey having discovered that this tract belongs to the Hubeishîyeh basin, which has its outfall into the sea, between the mouth of the Kasimîyeh and the city of Tyre.

Another geographical explanation is necessary before proceeding further. The basins of the Mediterranean Slope may be divided into two classes. Those of the first class are coterminous with the Jordan Basin. The second class basins are separated from the Jordan by the interposition of the upper parts of first class basins. This remark bears especially on the structure of the country which includes Tyre and Acre on its coast. In that part it will be found that only four consecutive basins in the interior divide the Kasimîyeh (or Litany) from the Mukutt'a (or Kishon) Basin. The four are—the Hubeishîyeh, Ezzîyeh, Kûrn, and N'amein; although there are about 30 distinct outfalls along the coast within this area. The Hubeishîyeh is reckoned among the first class basins, although it is divided from the Jordan Basin by the southern extension of the Kasimîyeh. But the latter is an exceptional case, for it conforms rather with the upper part of the first class basins of the Jordan's Western Slope, than with similar features on the Mediterranean Slope.* The second class

* See pages 186, 187.

basins in some cases are nearly as important as those of the first class, while others scarcely deserve notice. The following are of the second class between Tyre and Acre, viz.:— Wady el 'Akkâb, Wady Shema, Wady ez Zerka, Wady Kerkera, Nahr Mefshukh, Wady el Majnuneh, and Nahr Semeirîyeh. The consecutive upper portions of the first class basins, constitute an elevated plateau, between the water-parting range, and an irregular, ill-defined, outer range, which gives rise to the streams of the second class basins, while it is intersected by the gorges through which the streams of the plateau descend to the sea.

The Basin of Wady el Hubeishîyeh.

This basin is in contact with the Kasimîyeh Basin from the sea eastward and southward to a mile west of Beit Yahûn; thence to Kefrah, it joins the basin of Wady el Ezzîyeh; and from Kefrah to the sea, it is bordered by the basin of Wady el 'Akkâb, and smaller channels near Tyre.

Wady el Hubeishîyeh was named Wady el Mezra'ah, by Dr. Robinson and others. Mezra'ah is a place on the Wady Ashûr, the name of the middle part of this wady. Scarcely anything beyond the name is found in Dr. Robinson's map. The map to Mons. V. Guerin's "Galilee" is more ample in detail, but the only name given to this wady in it, is O. Achour, the equivalent in French to Wady Ashûr. Much more complete in this part is Lieutenant Van de Velde's map, and also the map of the Holy Land in Dr. William Smith's Ancient Atlas. But in those maps as in Mons. Guerin's and others, the Wady Humranîyeh (which is the northern branch of el Hubeishîyeh) is unnamed and misrepresented, its upper part being made tributary to Nahr Kasimîyeh. The addition of the upper Humranîyeh to the basin of el Hubeishîyeh extends the basin as far north as Sarifa (Therifeh in some maps).

The Basin of Wady el Hubeishîyeh is now found to be drained by two main channels, of which the Wady el

Hubeishîyeh includes the outlet into the sea on the north of Tyre, together with the central and southern parts of the basin. The northern part is watered by the Wady el Humranîyeh and its tributaries.

The Hubeishîyeh is so named from the sea to the village of Jilu, where it receives the drainage of the chief part of the centre of the basin, which lies between Kefr Dunin and Mezr'ah. This part is further distinguished by the villages of Dibâl, Teir Zinbeh, Kefr Dunin, Juweiya, Mujeidil, and Deir Kantâra. It is drained by four parallel valleys and their branches, which unite on the east of Jilu.

From Jilu upwards, and beyond its confluence with the Wady el Ma, the main channel is called Wady Ashûr. About two miles south of Jilu, the Ashûr receives a small affluent from the east of Mezr'ah, which completes the central part of this basin.

The Wady Ashûr is prolonged southward beyond Deir Amîs, so as to make it the recipient of Wady el Ma. Still it is the latter wady which should be considered as the main stream or channel. For south of Deir Amîs, the Ashûr derives its supply from two channels and their branches which embrace the drainage between Kh. el Jelameh (alt. 1,560 feet), and Kh. el Yadhûn (alt. 2,512 feet), including the village of Kefrah. This area may be regarded as an equilateral triangle with each of its sides about three miles in length.

The Wady el Ma is the outlet of an area much more considerable and remarkable. To its outlet at Deir Amîs, the Wady el Ma descends from the east, through a deep gorge with a very winding course, which leads up to the village of Safed el Battikh (alt. 2,220 feet). The direct distance is about five miles, increased to seven miles by the windings. This represents the length of the southern division of the basin, its breadth varying from two to four miles.

At Safed el Battikh, the main channel comes down from the north-east corner of a parallelogram, extending in length about six miles south-westward, and about four miles in breadth. A mountain spur, projected from Kh. el Yadhûn (alt. 2,512 feet) is the northern limit of this parallelogram, all the drainage of which descends to the channel which skirts the southern base of the spur, and runs in a north-east direction to Safed el Battikh, where it bends round the end of the spur, and follows its northern base to the Wady el Ma, in a direction parallel, but contrary to its higher course, until it is deflected by a great bend to the north on its way to Deir Amîs. On the summit of the eastern end of the spur is the Crusader's Castle of Tibnin or Toron.

The northern division of this basin is drained by the Wady el Humranîyeh. It has on the north the waterparting dividing it from the Kasimîyeh; and its division from the central part of the basin on the south, may be defined by a line from the outfall through M'arakah and Kefr Dunin. It is nowhere so much as three miles wide. The course of the main channel is north-westerly until it approaches its outfall, when it turns to the south-west. The head-waters are collected in two parallel valleys, dividing the villages of Silah, Baflei, and Neffakiyeh and uniting in the main channel below Baflei.

THE BASIN OF WADY EL EZZÎYEH, AND MINOR BASINS ON THE NORTH.

This basin joins that of el Hubeishîyeh in its upper part, although towards the sea they are divided by no less than six minor basins with distinct outfalls. Of these Wady el 'Akkâb and Wady Nettarah are the chief.

The Palestine Exploration Survey has revealed much inaccuracy in the former delineation of the minor basins. The wady which passes the villages of Siddikin and Kana was called Wady Shemaliyeh, and was supposed to reach the sea on the north of Tyre. This wady is now found to be the upper part of Wady el 'Akkâb, which makes a sharp bend at

'Ain Ibal, (formerly 'Ain Baal) and enters the sea at er Rusheidîyeh, within three miles south of Tyre.

Next on the south is another minor basin beginning near Yater, on the south-west of Kefrah. Its main channel, in the upper part, bears the name of Wady Nettarah. It runs north-west to el-Kuneiseh, where it receives a small branch from the village of Kana, and there bends at right angles on its way to Deir Kanun, Shema'aiyeh, and the coast, which it reaches at Ras el 'Ain. The relation of this wady to the places connected with it in former maps is materially altered by the new survey.

The basin of Wady el Ezzîyeh is quite confined to the banks of the wady for three miles above its outfall into the sea. It becomes about two miles wide where the main channel is sunk in a deep gorge below the village of Zubkin, which stands on the waterparting between the Ezzîyeh and the minor basins on the north. East of the mountain of Kh. Belat (alt. 2,467 feet) and crossing the villages of Beit Lif and Ramia, the basin has a width of about four miles. Further east the basin attains to its greatest width, or ten miles, between the waterparting at Harîs on the north, and at Sasa on the south. From the mouth of Wady el Ezzîyeh on the west to Marûn er Ras (alt. 3,083 feet) on the east, the distance in a straight line is about 16 miles. At Marûn er Ras the waterpartings of the Ezzîyeh, Kasimîyeh, and Jordan basins meet.

Between the sea and the neighbourhood of Kefrah, the basin of Wady el Ezzîyeh is coterminous with the minor basins of Leileh, Nettarah, and 'Akkâb. Near the villages of Kefrah, Harîs, and Haddâtha, it meets the Hubeishîyeh basin. From thence to Marûn er Ras it runs south-easterly with the Kasimîyeh basin. From Marûn er Ras to Sasa its course is south-westerly along the waterparting of the Jordan, or more particularly, the tributary basin of the Jordan which has its outfall through Wady Hindaj into Lake Huleh. Near Sasa is the southernmost limit of the Ezzîyeh basin; and here the waterparting bends at a right angle to the north-west, crossing

Jebel Adâther (alt. 3,300 feet) which divides the first class basins of Wady el Ezzîyeh and Wady el Kûrn. From the foot of Jebel Adâther, the waterparting runs with second class basins, beginning with the important Wady Kerkera; north of which it meets a series of small basins confined in a triangular space between the lower part of Wady el Ezzîyeh and the mountain range of Jebel el Mushakkah, so well known by its termination on the coast in Ras en Nakûrah. This group of minor basins has undergone much rectification from the Palestine Exploration Survey. The largest of them are Wady Shemá and Wady ez Zerka,* the latter rising in the mountain of Kh. Belât (alt. 2,467 feet), from which is a panoramic prospect of great extent and beauty. The Tell Belât (alt. 2,020 feet), is a distinct summit to the south in the midst of the Wady Kerkera basin.

Among the dubious questions set at rest by the Palestine Exploration Survey, none is more striking than the topography of the Ezzîyeh basin. Robinson's map throws but little light upon it. Van de Velde's map and the "Holy Land" edited by Dr. George Grove for Dr. William Smith's Ancient Atlas, are remarkable approximations to the truth. But the latest map in Mons. V. Guerin's elaborate "Description de la Galilée" erroneously throws the whole of the southern part of the upper Ezzîyeh basin—from Kh. Shelabun (Guerin's Kh. Cha'laboun) to Kh. el Kurah (Guerin's Kh. Koura)—into the Kerkera basin. So also does Lieutenant Van de Velde.

The Basin of Wady Kerkera.

Although this basin is only of the second class, being divided from that of the Jordan by the southern part of the upper plateau of the Ezzîyeh, it is of some extent and not devoid of natural features. These however seem to have escaped the travellers who have visited this region, among

---

* The Wady Hâmûl of Dr. Robinson, "Bib. Res.," iii, 65.

whom are Dr. Tristram, who camped at el Basseh, but was drawn off from the smoother features of Kerkera, to the bold gorge of Wady el Kûrn and its fine ruined castle which he has so well depicted. Mons. Guerin has reported on several of the ruined sites and• villages of the Kerkera but he had no eye for the river basins, or for the mountainous and other natural features, except in a picturesque point of view, his attention being chiefly attracted by the present inhabitants, and the remnants left by their far more numerous and wealthy predecessors, who have passed away, leaving only abundant proofs of the natural capacity of the country at large to support a much larger population than the present. These smiling fields and pastures; woodlands, orchards and gardens; picturesque hills and valleys; amid high mountains and deep, precipitous gorges; would naturally swarm with people, if good government could be secured. In this desideratum all the manufacturing and trading nations of the earth have a common interest, no less than the sovereign and people of the localities immediately concerned, for whatever adds to the productiveness of the soil promotes consumption in general, and enlarges the demand for every article that adds to the comforts of life.

Dr. Robinson ignores Wady Kerkera altogether, and generally confuses the topography of this tract.* For it is the Kerkera, and not Wady el Kûrn, which drains the southern slope of the range that terminates in Ras en Nâkûrah. And as for Teirshiha, it is quite within the basin of Nahr Mefshukh (Mabshuk), instead of being on the southern side of Wady el Kûrn; although no doubt it may appear to be so from the distance at which Dr. Robinson made his observation. In tracing the Wady el Kûrn up to Jebel Jurmuk, Dr. Robinson is quite right as he generally is in dealing with the broader aspects of his subject. No one can be a greater admirer of Dr. Robinson's geographical genius than the present writer, and these remarks, far from

* "Bib. Res.," vol. iii, 66.

being meant to be personally invidious, are only recorded as proofs of the value of the Palestine Exploration Survey, and the impossibility of placing dependence on observations of a less exhaustive character.

It remains to be said that the Kerkera basin stretches southward from the summit of Jebel Mushakkah over a distance of three or four miles, its southern margin passing through Kh. 'Abdeh, Kh. Jelîl, and Fassûtah. Tel Belât (alt. 2,020 feet) rises in the midst of the basin. The head of the basin on the east, runs with the southern part of the Upper Ezzîyeh, between Kh. Belât (alt. 2,467 feet), and Jebel Adâther (alt. 3,300 feet). Its length is about 15 miles.

THE BASIN OF WADY EL KÛRN.

This is the Nahr Herdawil of some writers. Although, at its outfall into the sea, this basin is separated from the outfall of Wady el Ezzîyeh, by several minor basins and the great headlands of Ras en Nakûra and Ras el Abyad, these two basins meet together in the highlands at Jebel Adâther, (alt. 3,300 feet), and by their junction divide the basins of Wady Kerkera and the Jordan, both of which also approach Jebel Adâther on the north-west and south-east respectively.

Between the sea and the north-western roots of Jebel Adâther, the basin of Wady el Kûrn is bounded by that of Wady Kerkera. Next to Jebel Adâthar and the Ezzîyeh basin, the Kûrn basin falls in contact with three great divisions of the Jordan basin, which have their outfalls: (1) by Wady Hindaj into Lake Huleh; (2) by Wady Amûd, which drains Safed into the Sea of Galilee; (3) by Wady er Rubbudiyeh, which like Wady Amûd reaches the Sea of Galilee through the Ghuweir or Plain of Gennesaret. This section of the boundary of the basin is the source of its head waters, and encircles them by mountains of the greatest height in Galilee, for south-east of Jebel Adâther is Jebel Jurmuk (alt.

3,934 feet), and further south is Jebelet el 'Arûs (alt. 3,520 feet). South of Beit Jenn, the waterparting of the Kûrn basin passes from contact with the Jordan basin, and meets for about three miles the north-eastern extremity of the great N'amein basin, which divides the Kûrn by a wide interval from the Mukutt'a basin, and empties itself into the sea on the south of Acre. Between the Kûrn and N'amein outfalls and basins, a series of minor basins occur, of which those of Wady el Majnuneh and Wady Mefshukh abut on Wady Kûrn. The villages of Seijûr in the N'amein basin, Kisra in the Majnuneh basin, and Teirshiha with Malia in the Mefshukh basin, mark the course of the Kûrn waterparting, till the basin contracts into the narrow gorge through which the stream descends from the highland, to the Maritime plain. The extensive ruins of Kulat el Kurein, the Crusaders' Castle of Montfort, dominate this gorge, and control the road which passes through it between the coast and the interior. Onward to the sea at ez Zib, the biblical Achzib, the basin remains confined to the banks of the stream.

The upper basin of the Kûrn is drained by two main branches which unite at Kh. Karhatha. The eastern branch descends by deep gorges north-westward along the western base of the mountain range which extends from Jebelet el Arûs (alt. 3,520 feet) to Jebel Jurmuk (alt. 3,934 feet); and it receives from the northern end of the same valley, but flowing in an opposite direction, a branch from Jebel Adâther (alt. 3,300 feet). At the foot of a long spur from Jebel Jurmuk the wady bends from the junction to the west, then to the north, and again westward to the junction at Kh. Karhatha, and onward to the sea.

The western branch rises near Beit Jenn and skirts the western waterparting of the basin up to Kh. Karhatha. It flows through the fertile plains of el Bukeiah, but it descends to the junction through a deep and rocky gorge between Suhmata and Teirshiha.

THE BASIN OF NAHR N'AMEIN AND THE MINOR BASINS BETWEEN IT AND WADY EL KŮRN.

This basin has its outfall into the sea at the northern end of the Bay of Acre and near the fortress. On the north it is coterminous with the minor basin of Nahr Semeirîyeh, which reaches the sea about four miles north of Acre, and forms one of the series of minor basins between the Nahr N'amein and Wady el Kûrn.

The Minor Basins.

The Semeirîyeh rises on the west of Deir el Asad in two branches, which are named Wady el Humeira and Wady el Jezzâzeh, and have the village of Julis between them, a third and shorter branch rises near the village of Yerka, and passing between Kefr Yasif and Abu Senan, takes the name of Wady Abu edh Dhaheb. The waterparting between this minor basin and that of Wady el Majnuneh on the north, is marked by the villages of Kuweikat, Amka, and Deir el Asad.

The Majnuneh basin extends four miles further into the interior than the preceding, and is bounded on the east by the upper basin of el Kûrn. Its principal branch rises near Kisra (alt. 2,520 feet) and skirts the southern edge of the basin. Another branch rises on the south of Teirshiha and runs close to the northern edge of the basin, under the imposing remains of Kulat Jiddin, the vast Castle Judin of the Crusaders, and said to have been repaired as late as 1750 by Sheikh Dhaher when he governed Galilee. Another wady drains the centre of this basin and comes down between the villages of Yanûh (alt. 2,200 feet) and Jett, and empties itself at Kh. Akrûsh.

The basin of Nahr Mefshukh lies between the Majnuneh and el Kûrn. At its head are the villages of Malia and Teirshiha, near which rise three branches which skirt the northern and southern edges, and the centre of the basin, respectively. Their junction takes place amidst a group of villages, including Kabry, et Tell, el Kahweh, el Ghabsîyeh, el Ferj, and el Jebakhanjy, amidst miles of gardens and orchards. The great fountain called el Kabry is here, and it supplies Acre with water through an aqueduct. The whole basin appears to be extremely fruitful and picturesque, and it retains many vestiges of antiquity.

The Semeirîyeh, Majnuneh, and Mefshukh basins, together with a very small but distinct tract between the outlets of Mefshukh and Kûrn, named Wady Sillik, complete the area surrounded by the basins of Wady el Kûrn and Nahr N'amein.

The Nahr N'amein Basin.

The Nahr N'amein is identified with the Belus or Pagida mentioned by Josephus and by Pliny. Its northern waterparting divides it from the Semeirîyeh, the Majnuneh, and the upper basin of Kûrn. This waterparting is, besides, a mountain range extending from the Jebelet el 'Arûs (alt. 3,520 feet) and presenting many bold and precipitous escarpments on its southern face, until it sinks into the maritime plain near Acre. The range forms the natural division between Upper and Lower Galilee. Except Jebelet el 'Arûs, the only height given along the range on the Palestine Exploration Survey is Jebel Heider (alt. 3,440 feet); until the mountain slopes downwards to the plain at el Judeiyideh (alt. 295 feet) and at el Mekr (alt. 191 feet). Captain Mansell's observations partly supply this want. He gives for Kûrn Hennawy (alt. 1,110 feet); for Mejd el Kerûm (alt. 1,294 feet); for the summit south of el Bukeiah, probably on the road to Seijûr (alt. 2,657 feet). As the Southern Range of Upper Galilee, it is further noticed in page 177.

The eastern waterparting of Nahr N'amein divides it from the Jordan basin, and from that division of it in particular which has its outfall by Wady er Rubudiyeh into the Sea of Galilee. It extends from Jebel Heider, near Beit Jenn to Ras Hazweh, near Arrabet el Buttauf. The direct length is

c

seven miles, but, in consequence of a bold projection of the Jordan basin westward, the length along the dividing line is about 12 miles. The only altitudes observed on this line are at its extremities in Jebel Heider (alt. 3,440) and Ras Hazweh (alt. 1,781 feet). The heights of Jebel el Kummaneh and Jebel 'Abhariyeh are desiderata.

The southern waterparting of Nahr N'amein runs altogether with that of Nahr el Mukutt'a. The only altitude observed on this line is at Jebel ed Deidebeh (alt. 1,781 feet). Between Ras Hazweh and Khurbet Jefât, the parting runs from east to west; from Khurbet Jefât it takes a south-western course till it closes on the permanent stream of Wady el Melek on the south of Shefa 'Amr; and from thence it proceeds north-westward to the Bay of Acre.

The greatest width of this basin is about ten miles north to south, through Shefa 'Amr; and its greatest length is about 19 miles.

Its principal channel is the Wady el Halzûn, called also higher up, Wady Sh'aib. It is the recipient of two head-waters which come respectively from Jebel Heider on the north-east, and Ras Hazweh on the south-east. Another affluent is Wady esh Shâghûr which joins it from el Baneh on the north. It has several tributaries along its left bank or from the south, among which the chief are the Wady el Balât which passes Kâbûl; and a smaller one rising near Tumrah and passing er Rueis.

North of Wady el Halzûn, is Wady el Waziyeh, which, with the affluent of the Halzûn rising near el Baneh, successively defines the southern base of the mountain range between Upper and Lower Galilee. These wadys are traversed by the principal route between Acre, Safed, and the noted passage of the Jordan at Jisr Benat Yakûb.

South of Wady Halzûn, is the Wady 'Abellîn, which rises on the west of Kh. Jefât, passes Kaukab and Kh. 'Abellîn and falls into the swamp of Nahr N'amein. Parallel with Wady Abellîn on the south-west, are three small wadys, of which the first descends from 'Abellîn; the second comes from Tell Saraj Alauneh and passes Shefa 'Amr; while the third skirts the waterparting up to Jidru, and then falls with the others into the end of the N'amein Swamp. The N'amein basin forms the north-western part of Lower Galilee. The south-western part is the Basin of the Mukutt'a and the whole of the eastern part is in the Jordan basin. The delineation of this important part of the country has undergone great improvement from the Palestine Exploration Survey, especially along the line of communication between Acre, Safed, and the East. The Halzûn also is now found to proceed straight into the swamp, instead of making a great bend to the south to join the 'Abellîn, from which it is really quite separate. The relation of the plains of Rameh and Sukhnin, and the correct position of the confluence of their main channels is now made intelligible. The proper allocation of the Valley of Jipthah-el may now be discussed, with a knowledge of local conditions that leaves nothing to be desired.

The Basin of Nahr el Mukutt'a (Kishon).

"The great battle-field of Jewish History and the chief scene of Our Lord's ministrations," to use Dean Stanley's impressive words, are embraced in this basin and its counterslope. The defeats of Sisera, Saul, and Ahab; Elijah's conflict with the priests of Baal; the deadly wounding of King Josiah; the Saviour's home at Nazareth; the marriage at Cana of Galilee; the resurrection of the widow's son at Nain; all occurred here. The great plains of Esdraelon and the lesser plains of Buttauf and Toran, the mounts of Carmel and Gilboa, and the highlands of Nazareth, characterise the scene.

The northern boundary of the basin, dividing it from Nahr N'amein, and extending from the Bay of Acre, to Ras Hazweh has been already described. The eastern boundary reaches from Ras Hazweh along the waterparting of the Jordan to Tannîn (alt. 1,460 feet). The direct distance is 30 miles; but following the windings the length is 50 miles. Several of the western affluents of the Jordan descend from the border of the Mukutt'a basin. Four of them occur between Ras Hazweh and Neby Duhy (alt. 1,690 feet). Between Neby Duhy and Sheikh Barkan (alt. 1,698 feet) is the Nahr Jalûd or the Valley of Jezreel, which falls into the Jordan below Beisân. Three more outfalls into the Jordan complete this series up to Tannîn.

From Tannîn to the sea, the waterparting takes a winding north-westerly course. As far as Mûsmûs road, the Mukutt'a is contiguous with the maritime basin of Nahr el Mefjir, formerly named Nahr Akhdar. The Nahr el Mefjir has a common waterparting with the Jordan, and is therefore of the first class. From Musmus road north-westward, the waterparting divides the Mukutt'a from the second class basins of Nahr ez Zerka, and Nahr ed Dufleh, the latter ending at Wady el Milh. Here the waterparting ascends the main ridge of Mount Carmel, which it pursues up to the deflection of the ridge road down towards Haifa, and following this descending route it meets the coast on the east of the town.

The Mukutt'a River falls into the southern part of the Bay of Acre near Haifa. About four miles from the bay, it is joined by the Wady el Melek, near the village of el Harbaj. The Wady el Melek is the main drain of the northern part of the basin. Its most distant sources are at Ailbun, Nimrin, and Lubieh, including the plains of Buttauf and Toron, and the hills as far south as Neby Sain (alt. 1,602 feet), close to Nazareth, from whence the southern boundary of the affluents of Wady el Melek may be traced by Ailut to Beit Lahm, Umm el 'Amed (alt. 643 feet), and Harbaj. The eastern part of the Plain of Buttauf has usually no outfall beyond its own swamp, which dries up in summer; but judging from Dr. Thomson's account of his passage from Rummaneh to Kana,* across 'a spongy morass, the overflow in floods probably reaches Wady Rummaneh. The River Mukutt'a itself drains the central and southern divisions of its basin. Its course is from south-east to north-west, the distance between its farthest head

* "The Land and the Book," p. 426.

near the village of Jelbon, and its outfall at Haifa, being 22 miles in a direct line without reckoning the windings.

The separation between the central and southern divisions may be identified with the highway between Mûsmûs, el Lejjûn, and el 'Afûleh. For south of this road the basin is prolonged between Mount Gilboa, and Mount Ephraim, and its drainage passes down through the intermediate plain by four main channels with numerous branches, all of which unite in one outfall before crossing the road. This distinct hydrographical area forms the southern division of the basin.

Between el 'Afûleh and Zerin, the waterparting marks the separation of the Plain of Esdraelon (Merj Ibn 'Amir) and the head of the broad valley of Jezreel (Nahr Jalûd), which here begins its steep descent to the Jordan.

The four main channels of this southern part of the basin have their sources (1) between Zerin (Jezreel) and Ras esh Sheiban;—(2) between Ras esh Sheiban, and Kh. el Medeka-kin, including a considerable extent of waterparting which may be traced through Jebel Abu Madwar (alt. 1,648 feet),—Jelkamus (alt. 1,308 feet),—Tannin (alt. 1,460 feet),—and Kh. Umm el Butm;—(3) between Kh. el Medekakin and Sh. Iskander (alt. 1,699 feet);—(4) between Sh. Iskander and el Mesheirfeh near Mûsmûs. The final confluence of these channels occurs close to the crossing of the boundary road, at an altitude of only 203 feet.

Two affluents of the Mukutt'a, forming the head of the central division, descend in opposite directions from the north-east and south-west respectively, and join the main stream from the southern division near Ludd. Their confluence is about 2 miles below the boundary of the divisions, and its altitude is 181 feet. The north-eastern affluent rises at et Tireh on the north of Iksal, the biblical Chesulloth, and east of Nazareth. It receives a branch from Neby Duhy (alt. 1,690 feet) and from Nein, the biblical Nain (alt. 744 feet). The south-western affluent descends from an alt. of 1,290 feet through the Wady es Sitt and waters the village of el Lejjun (alt. 403 feet). Three noticeable affluents descend to the right bank of the main stream from the highlands of Nazareth; and three smaller ones come from the oak-clad hills which are interposed between the Plain of Esdraelon and the Maritime Plain of Acre. These forest hills rise to altitudes of 600 feet, and the Mukutt'a forces its way to the sea through a gorge which divides them from Mount Carmel (alt. 1,810 feet).

The tributaries on the left bank are more numerous than noteworthy. West of el Lejjun a spur descends from the waterparting (alt. 1,290 feet) to the village of Ludd (alt. 275 feet) and the Mukutt'a River (alt. 150 feet). On the slopes of this spur are found remains of the ancient city of Megiddo. On the east it sends off a perennial branch to the Lejjun tributary. On the west its numerous rivulets are spread out from Buseileh to Abu Shusheh and reach the Mukutt'a by four outfalls.

Northward of the spur, the steep descent of the highlands of Ephraim, terminates in the plain about midway between the waterparting and the course of the Mukutt'a, until the foot of Mount Carmel is reached, when the Mukutt'a dives into the gorge which leads it to the maritime plain and the Bay of Acre. Between Abu Shusheh and Mount Carmel three notable affluents on this side of the Mukutt'a are—(1) a perennial stream which descends from the village of Jarah (alt. 834 feet);—(2) another perennial stream which waters Kh. er Rihâneh and Kh. el Farriyeh;—and (3) the Wady el Milh, which drains the southern end of Mount Carmel, including el Mahrakah (alt. 1,687 feet) the place of Elijah's conflict with the priests of Baal, also the eastern face of Umm ez Zeinat, passing from the hills at the foot of Tell Keimûn (alt. 248 feet).

Mount Carmel, Eastern Slope.

The north-eastern face of Carmel with the watercourses which furrow it and descend to the Mukutt'a, and the pathways that surmount it from various points of the highway that skirts its base on both sides, are now so distinctly delineated by the Palestine Exploration Survey, as to lay bare its topographical features with a precision of detail, which the most attentive of previous travellers and surveyors could not have contemplated, and certainly did not achieve. As Mount Carmel is a special favourite and of limited extent, its treatment will be made more complete than other parts. From the foot of Tell Kaimun, paths ascend to the ridge of Carmel by the Wady el Milh, and also by el Mahrakah. The ridge road extends through Esfia to the convent, all along the summit. Northward up to Jelameh the watercourses of this face of Carmel are precipitous and there are no paths. At Jelameh, there is a zigzag track up to the village of Esfia on the ridge (alt. 1,742 feet). From Esfia the Wady esh Shomariyeh has an oblique and therefore easier drop into the Mukutt'a. The Wady abu Haiyeh is the next on the north, and descends from the south side of Jebel 'Akkara (alt. 1,715 feet). About a mile from Esfia, a path descends from the ridge over Jebel 'Akkara to the foot of the mountain at Belled esh Sheikh. The northern side of Jebel 'Akkara is drained by Khallet (Ravine) en Nury, which runs north-westwards for two miles in a valley parallel with the main ridge of Carmel, when it breaks through a gap in an outer ridge, and descends north-eastward, through Wady et Tabil, to the Ashlûl el Wawy, an affluent of Mukutt'a, the latter ceasing to be the direct recipient of the mountain wadys, north of the village of Yajur. The elevated valley or ravine of en Nury is prolonged for a third mile parallel with the ridge up to Ras ez Zelâka (alt. 1,535 feet). At Ras ez Zelâka a path descends from the ridge through this valley and Wady et Tabil to Belled esh Sheikh. The drainage of this prolongation is southeastward, in direct opposition to that of the other part of the valley, which it joins before entering the gap in the outer ridge, on its descent to Wady et Tabil.

Besides Wady et Tabil, the Ashlûl (cascades) el Wawy receives some short channels near the village of Belled esh Sheikh, and also the Wady Hawasah and another unnamed wady from the north side of Ras ez Zelâka. It becomes a permanent stream at 'Ain es Sadeh, and falls into the Mukutt'a about a mile from the coast.

The remainder of Carmel is independent of Mukutt'a. The Wady Abu Mudauwar, the Wady Rushmia, and some others have a common outfall in the Nahr Mantney, close to Haifa, and complete the drainage of this side up to a mile and a half of the convent. A path comes from the ridge at the head of Wady Mudauwar, and follows the edge of the wady to Haifa. Four more paths descend from the ridge to Haifa, at intervals between this point and the convent.

The Palestine Exploration Survey of the great basin of the Mukutt'a, well deserves to be remarked before leaving it, as a notable example of the surveyor's work; for although its general aspect was delineated in previous maps, rectifications and very numerous additions occur in every part of the new Survey.

MINOR BASINS WEST OF THE MUKUTT'A, INCLUDING WESTERN CARMEL.

South of the Mukutt'a, the next basin which extends back to the waterparting of the Jordan, is that of Nahr el Mefjir, the outfall into the Mediterranean Sea being near the ruins of Caesarea.

All to the north of the Mefjir basin, and between the Mukutt'a on the east, and the Mediterranean on the west is included in a triangular area, drained by numerous petty wadys with many independent outfalls. The western slope of Mount Carmel forms the northern part of this triangle, and recalls attention to this mountain, the account of which will now be completed. The chief of the channels which descend from the western side of Mount Carmel are,—(1) the Wady el 'Ain, which in its middle course runs parallel to the range, and, reaches the maritime plain at the village of et Tireh. (2) the Wady Fellah with its outfall near 'Athlît, which receives the Wady el Miftelah from a valley which runs parallel with the main range of Carmel for about four miles. The Wady

Fellah receives two affluents from the villages of Daliet el Kurmul (alt. 1,245 feet) and Ain Haud (alt. 357 feet). (3) The Wady el Mikteleh which rises on the south of Daliet el Kurmul and reaches the shore if not the sea near Khurbet Malhah. (4) The Wady el Matabin, which rises in Ras el Meshahir (alt. 1,510 feet), an outlying summit of Carmel on the east of el Mahrakah. It receives a tributary from Umm ez Zeinat, flows by Ijzim (alt. 387 feet) and reaches the plain near Kefr Lam and Surafend. Wady el Milh, and the Wady el Matabin, together define the southern base of Mount Carmel, and its separation from the lower region of Mount Ephraim. These Wadys are also traversed by a road which completes the communication round the bases of the mountain. The southern limit of Mount Carmel will again be discussed in connection with an account of the range of hills on the south of Nahr ed Dufleh.

Mount Carmel has been hitherto delineated as a single ridge; but the Palestine Exploration Survey has brought to light parallel ranges more or less developed on each flank of the main ridge. The central or waterparting ridge extends from the Convent (alt. 470 feet) through Ras ez Zelakah (alt. 1,535 feet) and the village of Esfia (alt. 1,742 feet), to a point east-south-east of Daliet el Kurmul, where four tracts meet, coming from Esfia, Daliet el Kurmul, Umm ez Zeinat, and Tell Keimûn. At this point is the head of Wady el Milh. Its altitude is not given, but about three-quarters of a mile northward is the highest observed point of the mountain (alt. 1,810 feet).

The parallel range westward of the waterparting is defined as follows:—From the junction of the four roads, the culminating ridge bends round to the westward, or west by north-half-north, and is traversed by the track to Daliet el Kurmul. But before reaching the village, the height of land turns northward and again westward to the summit of Ras Umm esh Shukf (alt. 1,607 feet). The northern flank of Mount Shukf continues this range to the gap of Wady Shellaleh, where Wady el Miftelah, after distinctly dividing the central and western range for about four miles, breaks through the western range to reach the sea at Athlit. From the gap of Wady Shellaleh, the western range skirts the Wady el Ain and descends to the plain at the village of et Tireh.

The eastern slope of Carmel presents a similar parallelism, chiefly displayed in the Khallet or Ravine of en Nury, which divides an outer ridge from the central one, for a distance of nearly three miles, between Ras ez Zelâkah and Jebel 'Akkara. Towards the south the outer ridge is found below Esfia, between Kh. esh Shelkîyeh and the Wady esh Shomariyeh. North of Ras ez Zelâkah, the outer ridge descends by Tell Abu Mudauwar, and encloses numerous channels between it and the central range. These channels unite before they break through the outer range to reach the sea through Nahr el Mantneh, near Haifa. Dr. John Wilson noticed this feature as exhibiting "a lateral gash in the hill, running in the direction of the promontory, which is of some magnitude. It is here that the best cultivated fields occur."*

THE BASINS OF NAHR ED DUFLEH AND NAHR EZ ZERKA.

There are two minor basins between the southern base of Mount Carmel and the first class basin of Nahr el Mefjir. These have their outlets by Nahr ed Dufleh, on the south of Tanturah, the biblical Dor of Manasseh; and by Nahr ez Zerka, which is identified with the Crocodile River† of Pliny, and the Shihor Libnath of the Bible. Together they drain the principal part of the hilly tract now called Belad er Ruhah.

The Basin of Nahr ed Dufleh.

Three permanent streams with tributary wadys drain the head of the Nahr ed Dufleh, between the villages of Daliet er Ruhah and Umm et Tût; and between Umm et Tût and

* "Lands of the Bible," ii, 242.
† See Macgregor's "Rob Roy on the Jordan," 6th edit., p. 387, *note*.

Shefeia, the Wady Madhy falls in from Kumbazeh, on the northern margin of the basin. Its length is about 11 miles, and its breadth about four miles. The hills which form the southern waterparting of this basin, and divide it from the Nahr ez Zerka, terminate on the west in the remarkable promontory of el Khashm, on the north of Caesarea. Here also the narrow plain of Tanturah* which extends along the western foot of Mount Carmel, and has a width of one mile and a half between el Khashm and the sea, suddenly expands into the famous Plain of Sharon, which attains a width of ten miles, and will be further noticed hereafter.

At the Zerka and el Khashm, the summit of the receding hills extends eastward in an arc behind Subbarin and Kefrein, to Umm el Fahm and Kefreireh, where the range sinks down to the Sahel or Plain of 'Arrâbeh, and the Wady Selhab in the Mefjir basin. The hills on the east of the plain are offshoots of Mount Gilboa; and those on the south culminate in the range of Sheikh Beiazid (Alt. 2,375 feet), on the north of Samaria, and a portion of the southern waterparting of the Mefjir basin.

The eastern waterpartings of the basins of Nahr ed Dufleh and Nahr ez Zerka, form a low saddle, which, with those basins and the Wady el Matabin on the north, as well as Wady el Mihl and the wadys of the perennial affluents of the Mukutt'a rising at Kh. er Ruhaneh and Jarah, occupy a depression of the hills called Belad er Ruhah, dividing Mount Carmel from the range of Jebel Sh. Iskander and Umm el Khataf, in the Basin of Nahr el Mefjir. See pp. 34, 222. The altitudes and other incidents lead to this conclusion, and thus is added orographic to the hydrographic evidence on this subject, which was first noticed in page 29.

The Basin of Nahr ez Zerka.

The basin of Nahr ez Zerka lies between the villages of Subarin on the north, Kefrein on the east, and Kefr Kara on

---

* So called by the present writer, in the absence of any previous name.

the south. Its principal wady with many small tributaries, drains the northern and eastern margins of the basin. The southern part supplies two main Wadys, one skirting the margin of the basin, and the other passing the village of Kannir, their confluence being due east of Caesarea.

From Cape Carmel to the Nahr ez Zerka, the delineation of the country has been very much altered by the Palestine Exploration Survey. For example, Sindianeh, ez Zerghâniyeh (Surganiyeh), Kh. Khudeirah (Gudara), Kannir (Kanir), and el Bureij, were all included in the basin of Nahr el Akhdar (now Nahr el Mefjir); whereas they belong to the more northerly basin of the Zerka. Koteineh appeared on the northern edge of the Zerka basin; it is now found on the northern edge of the Nahr ed Dufleh. Formerly the Dufleh basin was represented by Nahr Keraji which was supposed to be near Iksim (Ijzim). That village is now found to be in the basin of Wady el Matabin.

It is only just to the Palestine Exploration Fund's Survey, and at the same time an assistance to students, to unravel the tangled geography which route surveys, casual observations, and compilations from such materials have hitherto supplied. Transpositions from one basin to another, like that which has just been described in the connection with the Zerka and Mefjir, will be found repeatedly necessary in pursuing this account.

THE BASINS OF NAHR EL MEFJIR, AND MERJ EL GHÛRÛK.

Under the name of Nahr Akhdar, this basin was supposed to have very narrow limits, the waterparting between it and the Mukutt'a basin, having been wrongly confined to two or three miles on either side of the village of Umm el Fahm. Its waterparting is now known to be coterminous, with that of the Mukutt'a from the north of Umm el Fahm (alt. 1,400 feet)* to Tannîn, a ruin on the south-east of Jenin, the direct distance exceeding 20 miles. At Tannîn, the Mefjir

* Sk. Iskander, near here, is 1,699 feet.

basin joins the basin of the Jordan, and continues with it up to Ras el 'Akra (alt. 2,230 feet), a further distance of six miles. At Ras el 'Akra, it is diverted from the basin of the Jordan by the remarkable inland basin of Merj el Ghûrûk, forming a parallelogram with an area of about 30 square miles, without any outlet. The Merj is enclosed on three sides by the Mefjir basin, which again comes in contact with the waterparting of the Jordan for about two miles, north of Yasid. The area now known to be covered by the Mefjir basin has been hitherto wrongly divided between the basins of Nahr Falik, now reduced to very narrow limits, Nahr Abu Zabura, and Nahr Akhdar.

The main channel of Nahr el Mefjir descends from the Jordan waterparting between Tannîn and Râs el 'Akra. It is there called Wady es Selhab, and by that name it passes Kubatieh, and the Plains of Dothân and 'Arrâbeh. It passes from the western end of the Plain of 'Arrâbeh in a southerly direction, till it receives a stream from the village of 'Arrâbeh, when it turns westward with a winding course, and takes the name of Wady el Ghamik. Here it receives several small branches from the range east of Saida, and is diverted to the north, and soon again to the west, taking the name of Wady Abu Nar, and skirting a group of hamlets bearing the generic name of Nuzlet; as Nuzlet esh Sherkiyeh, Nuzlet el Wasta, Nuzlet et Tinat, and the Nuzlet el Masfy is not far off. The Wady Abu Nar enters the Plain of Sharon, between Jett and Baka, and joins the Wady el Maleh, near el Mejdel. After the junction the Maleh proceeds north-westward, and becomes a permanent stream in Wady el Khudeirah and Nahr el Mefjir.

The Wady el Maleh is also the recipient of Wady er Roz, which skirts the southern waterparting. This Wady rises at Yasid (alt. 2,340 feet) between the waterpartings of the Jordan, Merj el Ghûrûk and Nahr Iskanderuneh, and bears the name of Wady Abu Kaslan. It passes the northern foot of the Sheikh Beiazid Mountains on which the villages of Jeba, Fendakumieh, Silet edh Dhater, and 'Attara are situated at altitudes between 1,200 and 1,573 feet. Near 'Attara a branch unites which drains the north-west border of Merj el Ghuruk (alt. 1,770 feet). The main channel continues along the southern edge of the basin, passing Deir el Ghusûn (alt. 827 feet), Attîl (alt. 385 feet), Jelameh and el Mejdel to the junction at Tell edh Dhrur. Above Deir el Ghusûn, a branch falls in from Bâtn en Nûry (alt. 1,660 feet), and the village of Ellar (alt. 760 feet). Jelameh on the edge of the plain is 260 feet high, while the wady at its foot in the plain is 60 feet.

Another considerable affluent joins Wady el Khudeirah from the north-east. Its principal branch is Wady Arah which skirts the north-western limit of the basin, and rises at Umm el Fahm, on the mountain of Sheikh Iskander (alt. 1,699 feet). The Wady Arah passes between Kefr Kara (alt. 451 feet) and Ararah (alt. 707 feet), and enters the Plain of Sharon between Kh. ez Zebadneh (alt. 320 feet) and Kerkûr (alt. 160 feet), joining Wady el Khudeirah near Tell edh Dhrur (alt. 152 feet). Its principal branch is the Wady el 'Asl, which rises near Yabid (alt. 1,220 feet), and passes Kuffin (alt. 460 feet), where it receives a tributary from Ferâsîn (alt. 727 feet). At the ruins of el Medhiâb, the Wady el 'Asl is joined by a wady from the north, which unites Wady Samantar, Wady el Yahmur, and Wady Râsein, all draining the western slope of the wooded upland of Umm el Khatâf. Below el Medhiâb, the wady is called Wady Bir Isîr, and it enters the plain near el Mes'ady (alt. 142 feet), and soon joins Wady el Khudeirah. See p. 218.

The whole of the Mefjr basin, with half of the Zerka on the north, or as far as Kannir, and a small part of the Iskanderuneh on the south, or up to Kakon, was supposed to have its outfall by Nahr Abu Zabura, which is now called Nahr Iskanderuneh. Dr. Robinson on his last journey in Samaria, travelled from Umm el Fahm to Ya'bud (Yâbid) along the waterparting. From Ya'bud he had a view of the Plain of Arrabeh and passed Ferasin to Abu Nar, on his way to Attil, thus crossing the basin at its widest part.*

* Robinson's "Biblical Researches," iii, 121–124.

With regret, it is observed that Monsieur V. Guerin in his "Description de la Palestine, Samarie" omits all that part of this basin, encircled by a line connecting Umm el Fahm, Tell edh Dhrur, Baka, Attil, Kefr Raay, Arrabeh, and Yabid; as well as a large tract to the north in the Zerka basin.

The Basins of Nahr Iskanderuneh and Nahr Falik.

The outfall which bears the first name on the Palestine Exploration Survey, enters the sea at Minet* Abu Zabûra, by which name the river also was called by former authorities. The basin is coterminous on the north with that of Nahr el Mefjir, and the waterparting on this side runs almost direct from Tell el Akhdar on the coast, to Yasid. There the Jordan basin is met, and continues along the eastern boundary, which runs from Yasid through Jebel Eslamiyeh (Mount Ebal) and Nablûs (Shechem), to the southern base of Jebel et Tur (Mount Gerizim). The southern waterparting divides this basin from that of the Nahr el 'Auja. It runs westward from Jebel et Tur, through Kh. Jafrûn, Ferâta, Kh. Askar, el Funduk, Kuryet Hajja and the road to Kefr 'Abbûsh, Kefr Zibad, Kefr Jemmâl, to Khurbet Nesha, round the Bir el Hanûtah, where it turns northward and parallel with the course of the Wady Kulunsaweh, to Umm Sur, where it turns seaward through Mukhalid.

All this, with part of the Mefjir basin on the north, pertained in former maps to the outfall of Nahr Falik; which is now found to be restricted to a part of the maritime plain between Mukhalid, Umm Sûr Miskeh, Kh. Sabieh, and el Jelil, a village on the coast, south of Arsuf, the site of Apollonia. The length of the Falik basin is about ten miles along the coast, and its greatest width towards the interior scarcely exceeds seven miles.

The entrance of the Iskanderuneh into the sea, is on the north-western angle of its oblong basin. A branch rising near Belâh (alt. 1,367 feet) drains the western part of its

* Minet signifies a harbour.

northern boundary and passing Kakon in the plain, joins the permanent stream not far from its outfall. The north-eastern part of the basin is drained by its most noted watercourse, the Wady esh Shair and its tributaries. This wady rises at the town of Nablûs and flows by Zawâta, where it receives a tributary from Asîret el Hatab (alt. 2,036 feet) at the northern base of Mount Ebal. Then it waters Deir Sharaf, and flows on to Ramin (alt. 1,095 feet) where another feeder unites, which rises at Yasid (alt. 2,340 feet) in the north-east angle of the basin, and washes the remains of Samaria (Sebustieh). Below Ramin it passes Anebta (alt. 545 feet) and takes the name of Wady Zeimer, up to its confluence with the Wady from Kulunsaweh, near Jisr (Bridge) el Maktabeh. On the south of Wady Zeimer are the villages of Kefr el Lebad, Dennabeh (alt. 506 feet), and Tul Keram (alt. 370 feet), below which the wady leaves the hills.

On the south of Wady esh Shair is another large watercourse which rises in Jebel et Tur (Mount Gerizim) and drains the south-eastern part of the basin as far west as el Funduk. This is the Wady et Tin, which after a very winding course leaves the hills between Irtah (alt. 340 feet) and Feron (alt. 581 feet) and joins the Wady Kulunsaweh at Burin. West of its source, it passes el Arak, Till (alt. 2,084 feet), and Surra (alt. 1,647 feet), where it flows due north, and then bends round to the west, and again turns to the north passing Kusein. At Kh. ed Deir, it meanders to the west-south-west to the foot of the village of Kur (alt. 1,257 feet). Above Kur, it receives a tributary from Kh. Jafrûn, Ferata (alt. 1,715 feet), Amatin, el Funduk (alt. 1,295 feet), Kuryet Hajja (alt. 1,300 feet), Kuryet Jit (alt. 1,686 feet), and Kefr Kaddum (alt. 1,206 feet), villages on the southern margin of the basin. This is the Wady el Haj Musa.

The south-western part of the basin is drained by the Wady Kulunsaweh, which comes from Bir el Hanutah (alt. 365 feet), in the south-west angle, and runs to the north along the western edge of the basin. The Wady en Naml joins it from the east, and rises between Kur and Kefr 'Abbûsh (alt. 1,365 feet), and passes Kefr Zibad (alt. 910 feet), Kefr Jemmal (alt. 642 feet), and Felamieh (alt. 434 feet),—all villages on the southern edge of the basin.

Two highways cross this basin at its eastern and western sides respectively. The eastern road passes through Nablûs, and connects Jerusalem with all parts on the north. The western road runs on the east of the Kulunsaweh, at the foot of the hills; and connects the ports of Jaffa and Acre. Other important routes radiate from Nablûs in every direction. Nevertheless the Palestine Exploration Survey proves that only the crudest notions previously existed concerning the topography of this important basin.

THE BASIN OF NAHR EL 'AUJA.

This is one of the most considerable basins on the western watershed of Palestine. It empties itself into the sea about five miles on the north of Jaffa. It becomes a permanent stream below the hills in four of its branches, two of which have their perennial sources in the Plain of Sharon, between Bir 'Adas and Jiljulieh; another, which is reckoned the principal, has immense fountains at Kulat Ras el 'Ain, rising at the foot of a mound, on the Ramleh road, near the village of el Mirr, where it forms a marshy tract covered with reeds and rushes. The fourth permanent branch has its fountain also near el Mirr, on the south of that village. These unite at Tell el Mukhmar, and form a river which is said to be nearly as large as the Jordan at Jericho, with a bluish tinge, dark, deep, usually sluggish, and hardly to be forded at any place. There is an old bridge over the stream, on the high road from Jaffa.*

The basin of Nahr el 'Auja has its northern waterparting along with the Nahr el Falik, and Nahr Iskanderuneh. The waterparting begins at the sea on the south of el Jelil, runs north-west with the Falik basin nearly up to et Tireh, and then follows the Iskanderuneh basin, up to the foot of

* Robinson, "Phys. Geog. Holy Land," 176, 177.

Jebel et Tur (Mount Gerizim), the details of which have been described. From Mount Gerizim it runs with the Jordan basin across the Sahel or Plain of el Mukhnah, and bends southward to Kh. el Kerûm (alt. 2,700 feet), Akrabeh (alt. 2,045 feet) and the curious basin of Merj Sia, which lies between heights of 2,835 feet on the north and 2,710 feet on the west. Here the waterparting takes a south-westerly course, crossing Tell Asûr (alt. 3,318 feet), the village of Beitin or Bethel (alt. 2,890 feet), and Bireh (alt. 2,820 feet). At this point is the junction of the basins of Nahr el 'Auja, Jordan, and Nahr Rubin. The southern waterparting running with the basin of Nahr Rubin forms a great arc, continuing south-westward from Bireh to Saris (alt. 2,347 feet), and Beit Mahsîr (alt. 1,790 feet), where it takes a north-west direction through Abu Shûsheh, skirting er Ramleh, Surafend, and the road to Jaffa on the west, till it turns towards the coast at the trigonometrical points with altitudes of 261 feet and 240 feet.

The intricate system of watercourses composing this basin is primarily divisible into a northern and southern division.

The northern division includes the four streams and their tributaries, which unite near Tell el Mukhmar. They drain the whole of the northern waterparting, and also the eastern side southwards to Bireh. The northern is separated from the southern division, by an interior or sub-waterparting, which runs from Bireh, to Abu Kûsh, first by the " Roman road" and then by the "ancient road." From Abu Kûsh, the dividing line proceeds close to Bir ez Zeit (alt. 2,665 feet), and round by the high road to Umm Suffah (alt. 1,997 feet), and so on through Deir en Nidham (alt. 1,934 feet), to Tibneh, Abûd (alt. 1,240 feet), and along the northern side of the road from Abûd to Deir Dakleh, whence it skirts the south side of Wady Sahury to Rantieh, and bending round north-westward to Khurbet Shaireh, passes over an altitude of 275 feet to the confluence at Jerisheh, where the two divisions unite.

The southern division has its outfall into the Nahr el

'Auja, by a permanent stream, which flows from the south, passing Jaffa and Sarôna. This stream is the recipient chiefly of the Wady Nusrah. Near the junction of Wady Nusrah, two small streams also join, which only drain a small portion of the southern waterparting included between Ramleh and the coast. The Wady Nusrah on the contrary is the recipient of all the rest of the drainage of the southern division of the basin.

The Watercourses of the Northern Division of el 'Auja Basin.

An account of the chief watercourses of the two divisions of this basin, is the next step towards understanding its topography. Of the four streams which unite near Tell el Mukhmar and receive all the watercourses of the northern division, the most northerly is the Wady Kalkilieh, which receives the drainage of the north-western edge of the basin as far east as el Funduk (alt. 1,295 feet). The Wady 'Azzûn is its most southern tributary, and rises between 'Azzûn and Kefr Lakif. Yet 'Azzûn was supposed by Dr. Robinson to mark the northern limit of the 'Auja basin. " Phys. Geog., Holy Land," p. 176.

The next of the four streams is the Wady Kanah, which has been identified with the Brook Kanah of the Bible, the boundary between the tribal territories of Ephraim and Manasseh. The Palestine Exploration Survey appears to throw remarkable light on this much contested subject. See the " Quarterly Statement," October, 1880. The Kanah is one of the two great wadys of this division which derive their origin from the edge of the Jordan basin. The Wady Kanah drains the north-eastern part of the division, from el Funduk, by Mount Gerizim and the Plain of Mukhnah, to Yanûn and Kh. el Kerûm (alt. 2,700 feet).

The third stream is the Wady Rabah, which rises at Haris (alt. 1,560 feet), a village identified by the Talmud with Timnath Heres, allotted to Joshua. This valley is only about half the length of the Kanah valley, and, like the Kalkiliéh, it does not extend back to the Jordan basin. This is the Wady Ribah of Dr. Robinson, "about half an hour north of Mejdel Yâba," which was supposed to be in continuation of the Wady rising on the edge of the Jordan basin at 'Akrabeh, and passing Kubalan ("Phys. Geog.," p. 100). The Palestine Exploration Survey connects Akrabeh and Kubilan with the fourth of this series.

The southern part of the northern division of the el 'Auja basin, is drained by the Wady Ballût and its feeders, the fourth of the series of streams uniting at Tell el Mukhmar. The heads of this Wady extend along the eastern waterparting of el 'Auja, in contact with the Jordan basin, from 'Akrabeh to the south of Beitin (Bethel), a direct distance of 15 miles. The principal channel runs from 'Akrabeh, south-westward, in a straight course to Kurawa ibn Zeid. It bears the names of Wady Bushanit and Wady Ishar; and all the drainage between 'Akrabeh and Bethel flows into it at the three following points—(1) Near Yetma, the Wady er Rumt falls in from Jurish, Kusra, and Jalud. (2) At Khurbet Keis, the Wady el Kub unites an intricate system of valleys as follows:— From Eh Sawieh, a channel descends south-westward to el Lubban, and joins the Wady el Kub at Abwein. At Lubban this channel receives the Wady Seilun with branches from Kuriyut, Shiloh, and Turmus Aya. Between Lubban and Abwein another wady falls in from Sinjil; and at Abwein another descends from Jiljilia. (3) At Karawa ibn Zeid, a long and winding wady comes down from the eastern waterparting which it drains from Sinjil to Beitin, as follows:— A series of valleys rising between Beitin and Abu Kûsh unite at Ain Sinia, and run on northward to 'Attâra. At the east of 'Attâra, Wady en Nimr falls in from Tell Asûr on the southeast, after receiving a wady traversed by the Jerusalem Road from Kh. Kefr Ana on the south; and also the Wady el Jib from the north-east near Sinjil. From 'Attâra the main wady proceeds to Ajul, Deir es Sudan, and Karawa ibn Zeid.

From Kurawa, the Wady Ballût zigzags westward to el Kefr (alt. 1,290 feet), Deir Ballût (alt. 895 feet), and Mejdel

Yaba (alt. 495 feet), where it leaves the hills, and passes north-westward to its junction with the Nahr el 'Auja.

Below Kurawa, the Wady Reiya takes up the drainage of the southern edge of the division, rising on the south of Neby Saleh (alt. 1,866 feet), and running on to 'Abûd (alt. 1,240 feet), to join the Ballût at 'Ain Sarina.

West of 'Ain Sarina, the Wady Suhary rises, and skirts the southern edge of the division among the hills, entering the plain between Kuleh and Rantieh; where the wady turns to the north and north-west, by Neby Tari and Fejja, to join the Ballût on the east of Mulebbis.

On the right bank, the Ballut receives only one notable valley, which rises between Merda and Ishaka, has branches from Selfit and Furkhah, and falls into the Ballût on the south of Berukin.

The Watercourses of the Southern Division of El 'Auja Basin.

At Kefr Ana all the watercourses of this division unite in Wady Nusrah, which runs on by Ibn Ibrâk, Selmeh, and Sarona, to the Nahr el 'Auja at Jerisheh, the junction being only about a mile and a half from the sea.

Between Selmeh and Rantieh, there are three small but distinct tributaries to Wady Nusrah. They come from the north-western edge of the division, and have their junctions at Selmeh, Ibn Ibrâk, and Kefr 'Âna, respectively.

Between Rantieh and a point on the east of 'Abûd (alt. 1,240 feet), a combination of wadys arises, which unite at el Khurab, and run on to Kefr 'Âna. The first descends from the edge of the division, on the west of ed Diurah, runs under Kh. el Muntar and Deir Alla (alt. 675 feet), and enters the plain to reach et Tireh and el Khurab. The second rises in the Khallet (ravine) es Salib, which skirts the northern edge of the division between Abûd and el Lubban, where it bends round sharply to the south-west, passing in a gorge below Rentis (alt. 685 feet), to Khurbet 'Azzâz. Here it receives a tributary coming from Abûd, and from Deir Abu

Meshâl (alt. 556 feet) and then zigzags south-westward to Beit Nabâla (alt. 260 feet), where it receives the Wady Shâhin from Shukbah (alt. 1,058 feet), and Kibbiah (alt. 840 feet), and enters the plain, through which it proceeds by Deir Tureif and Kh. er Râs, to the junctions at el Kurâb and Kefr 'Ana.

The next outfall at Kefr 'Ana receives the greater part of the drainage of this division. It combines two separate drainage systems, which respectively concentrate in the Wady esh Shellal and the Wady Ludd, and these unite midway between Kefr 'Ana and the ancient town of Ludd. The Wady esh Shellal was formerly known as Wady Budrus.

The interior or sub-waterparting between the systems of Wady esh Shellal and Wady Ludd, is traceable from Jindas on the north of Ludd, over a trigonometrical station of 222 feet in height, to Deir Abu Selâmeh, Kh. Midieh (the remains of Modin of the Maccabees), and by the highroad to Shilta and Kefr Rût (alt. 1,290 feet), where it skirts Kh. Fa'aûsh and Kh. ed Dirish, and reaches Beit Ur et Tahta or Lower Beth Horon (alt. 1,910 feet). Here it ascends the pass to Beit 'Ur el Fôka or Upper Beth Horon (alt. 1,022 feet), and follows the high road between Jaffa and Jerusalem, till it reaches the top of the descent into Wady el Askar, where this sub-waterparting joins the main waterparting, between the Basins of Nahr el Auja and Nahr Rubin.

The Watercourses falling into Wády Shellal (Budrus).

The northernmost of these channels rises at Kh. Bir ez Zeit (alt. 2,665 feet) and runs north-westward alongside of the sub-waterparting and the high-road to Jaffa, nearly to Neby Saleh (alt. 1,866 feet). Here the channel bends about considerably in advancing westward, through a valley having Deir en Nidham (alt. 1,934 feet), Tibneh, Deir Abu Meshal (alt. 556 feet), and Shukbah (alt. 1,058 feet) on the north; while on the south are the villages of Kubar (alt. 2,021 feet), Beit Ello (alt. 1,797 feet), Jemmâleh (alt. 1,694 feet), and Shebtin (alt. 904 feet).

At Shebtin, another winding valley falls in from sources between Bîr ez Zeit and Batn Harâsheh (alt. 2,490 feet). This valley is further bounded on the south by Janiah (alt. 1,813 feet), Ras Kerker, Khurbetha Ibn Harith, and Deir el Khuddis (alt. 1,264 feet). Below Shebtin, the Wady takes the name of Wady en Natuf, bends round to the south-west, and reaches the foot of N'âlin. The village is 860 feet, and the wady below is 500 feet above the sea. West of N'âlin, the Wady en Natuf is joined by the Wady Malâkeh, a considerable affluent from the southernmost part of the Shellal system. The Wady Malâkeh has several important branches. as follows:—(1) The Wady Hamis is a permanent stream derived from fountains on the north of Ram Allah (alt. 2,850 feet), and joined by branches from the waterparting between Bireh and Abu Kûsh. (2) The Wady Kelb has a parallel course on the south of Wady Hamis, and also comes from Ram Allah; these unite between Janiah and Deir Ibzia, and take the name of Wady Dilbeh, which lower down is Wady Shamy. The Wady Dilbeh or Shamy passes Kefr Namah (alt. 1,483 feet), and Bel'ain, and enters the Wady Malâkeh, which at the same point receives (3) the Wady Ain 'Arik from the west of Ram Allah. The Wady Ain 'Arik is the recipient of (4) the Wady el Imeish, which runs parallel with the Jaffa-Jerusalem road, and after receiving the Wady es Sunt from Beitûnia, passes the Beth Horons, on its way to the junction with Wady Ain 'Arik. Where the Wady Malâkeh receives the Wady Shamy and the Wady Ain 'Arik, the Wady el Muslih (5) also falls in, from Lower Beth Horon and Suffa. From this confluence (alt. 713 feet) the Wady Malâkeh finally proceeds by Shilta to Midieh, and the junction with Wady en Natuf on the west of Nalin. Below the confluence the single channel of Wady esh Shellal runs westward to join Wady Ludd.

The Watercourses of Wady Ludd.

Two branches unite on the eastern side of the town of Ludd. The western branch is the Wady Harir, which collects the drainage of the southern waterparting from the neighbourhood of Abu Shusheh to Er Ramleh. The eastern branch is the outlet of the Wady 'Atallah and Wady Aly with their tributaries, which drain the country between the southern boundary of the Shellal or Budrus system, and the southern waterparting of the el 'Auja basin.

Four small wadys carry to the right bank of Wady 'Atallah, the drainage of a triangular space bounded by Deir Abu Selâmeh, Kh. Midieh, Shilta, and Kefr Rût (alt. 1,290 feet); where the limits of this space bend round southward by Bîr Maîn (alt. 940 feet), to Selbit, and the confluence of Wady Suweikeh with Wady 'Atallah. The village of Jimzu lies between two of these wadys on the north; the third is Wady Jâar, with the villages of Annabeh, Berfilya, El Burj, and Bir Maîn; the fourth is Wady Suweikeh in the midst of several ruined sites.

At El Kubâb on the 'Atallah, is the junction of Wady Selman (or Suleimân), which rises on the waterparting of the el 'Auja basin, at an altitude of 2,065 feet, on the west of el Jib, or of the plain around Gibeon. The ordinary camel route between Ramleh or Ludd and Jerusalem follows this valley. It has the villages of et Tireh and Kurbetha-ibn-esSeba on the north, with Beit Dukku and Beit Likia on the south.

The Wady Selman receives a branch on the north or right bank, from the south of the high road between the Beth Horons. Before the junction, the Wady el Mikteleh from Beit Sira (alt. 840 feet) joins the branch which falls into Wady Selman about a mile lower down at an altitude of 625 feet.

Below this junction the Wady el Burj joins the Selman on its left bank, on the north-west of Beit Nuba. It rises between el Kubeibeh (alt. 2,570 feet) and Beit Anan (alt. 2,070 feet), in the Khallet or ravine of el Kuta.

West of Beit Nuba (alt. 737 feet), the Selman receives, on its left bank, the Wady Mozarki, which rises at el Kubeibeh and passes Katanneh and Yalo (alt. 940 feet). At Yalo, a branch falls in, which rises near Kuryet el Enab, takes the name of Wady el Hai, and receives the Wady Khushkush.

Lastly, about three miles east of el Kubab, the Selman receives a small wady from the east of Amwas.

At el Kubab, above the junction of the Selman, the Wady 'Atallah takes the name of Wady Aly, and rises on the southernmòst portion of the waterparting of the el'Auja, near the village of Saris (alt. 2,347 feet). South-west of Latron (alt. 800 feet), the Wady Aly receives a feeder from the margin of the basin which has several heads on either side of Beit Mahsir (alt. 1,790 feet). On the north, near Deir Eyub (alt. 1,070 feet), the Wady Aly receives the Wady Alakah skirted by the road to Kuryet el Enab. The direct road between Ramleh and Jerusalem follows Wady Aly; but if less circuitous it is more difficult than the route by Wady Selman.

Dr. Robinson was under the impression that a branch of this system originated in the rugged chasm on the north of Ram Allah, and issued from the mountain north of the lower Beth Horon; but here he was at a loss to say whether it proceeded to Beit Nuba, or went on directly west to Wady Ludd ("Phys. Geog.," p. 102). This misconception was corrected by Lieutenant Van de Velde; and further improvements were represented in the map of the Holy Land prepared under Dr. Grove for Dr. Wm. Smith's Ancient Atlas. But the difference between such approximations and the actual topography, as it is delineated in the new maps, can only be duly appreciated after a study of the maps, for which these notes may afford some preparation.

THE BASIN OF NAHR RUBIN.

Between Jaffa and the town of Yebna (bib. Jabneel), the Nahr Rubin enters the sea, near the Neby Rubin, from which the name is derived. The basin of Nahr Rubin, with its central wady, follows the curvature that characterises the waterparting which divides it from the Nahr el 'Auja basin on the north. The eastern waterparting extends along the Jordan basin from Bireh to er Ram, Shafat, Jerusalem, Bethlehem, and Urtas, about three miles west of which, at a point south-west of el Khudr (alt. 2,832 feet), it joins the southern boundary. The southern waterparting dividing it from the larger basin of Nahr Sukereir passes from el Khudr along a highway towards Kefr Som and Beit Atab. South of Kefr Som, where the road bifurcates, the waterparting follows the south-western road towards Beit Netif (alt. 1,517 feet) along the northern edge of Wady el Werd. From Beit Netif, it passes to the north of Zakariya (alt. 940 feet); and with a bend to the north, proceeds westward to Mughullis, where it turns to the north again, and impinges on the Wady el Menâkh to the south of Jilia. Hence it goes westward to el Kheimeh (alt. 300 feet), passing Beshshit (alt. 197 feet), and Yebnah, to the coast at the Minet or Harbour of Rubin.

The Watercourses of Nahr Rubin Basin.

The Nahr Rubin appears in the Palestine Exploration Survey, as a permanent stream at a further distance from the coast than any other river in Palestine, except the Kasimîyeh; its perennial sources being traced up the Wady es Surar to the hills about Surah and Tibnah, made famous by Samson and the Patriarch Judah. Dr. Robinson, however, has recorded that in autumn the Nahr Rubin sometimes dries up.*

The head of the Nahr Rubin basin lies in a recess projected northward between the basins of the Jordan and el 'Auja. It contains the historical plateau of el Jib or Gibeon, which extends southward to the heads of the descending ravines and gorges of Wady el Ghurab, Wady es Surar or Ismain, and the "deep and rugged" Wady el Werd. This southern edge of the plateau,† is defined by the road from Jerusalem, which runs along the ridge on the south of Lifta, to the Wady es Surar on the north of Kulonieh, and thence

* Robinson's "Phys. Geog. H. Land," p. 177. † "Bib. Res.," iii., 159.

to the waterparting through Beit Surik, along the wady descending to W. es Surar, from that village. The western side of the recess extends along the waterparting from near Beit Surik, to the east of Biddu, Beit Izza (alt. 2,621 feet), and Beitunia (alt. 2,670 feet), to el Muntar, a trigonometrical station of the Palestine Exploration Survey with the altitude of 2,685 feet. It may be observed that the heads of Wady el Imeîsh and Wady Selman rise in the plain of el Jib, and are not separated from Wady ed Deir by any prominent elevation. From el Muntar the northern side passes through Ram Allah (alt. 2,850 feet) to Bireh (alt. 2,820 feet), the biblical Beeroth, a village on the Jerusalem-Nablus road. The eastern side passes along the edge of the Jordan basin, from Bireh by Kefr Akâb (alt. 2,740 feet), er Ram (alt. 2,600 feet), and Tell el Ful (alt. 2,754 feet), to the west of Jerusalem.

Two parallel watercourses with tributaries run through the plateau from north to south. The western is Wady ed Deir, which rises at Ram Allah, and "winding among low hills,"* has el Jib (alt. 2,535 feet), and Neby Samwil (alt. 2,935 feet) on its right bank. On this side the Wady Amir comes from near Biddu, and passes between el Jib and Neby Samwil.

The eastern watercourse rises as Wady Jillan near Bireh, and skirts the Jerusalem road in passing Kefr Akab and er Râm (alt. 2,600 feet) on its left bank. South of er Râm, the wady takes the name of ed Dumm, and turns to the southwest, to meet the western branch, on the south of the village of Beit Hannina, which lies between the wadys. The villages of Tell en Nasbeh, Rafat, Kulundia, Jedireh, and Bir Nebala, are also situated in a "beautiful plain"† between the two wadys.

Below the confluence the wady is called after the village of Beit Hannina, and it runs on southward for about two miles to Lifta, where it bends to the westward for a mile and a half, receiving the Wady el Âbbeideh from Neby Samwil and

* Robinson's "Bib. Res.," i., 454. † *Ibid.*

Beit Iksa. The Wady Buwai from Beit Surik, joins the Wady Beit Hannina where it bends round on a southerly course to the village of Kulonieh.

Having some time since paid close attention to the cartography of this interesting plateau, in constructing the map of the environs of Jerusalem in Dr. Wm. Smith's Ancient Atlas, the present writer may be allowed to express the gratitude of a geographer for the light which the Palestine Exploration Survey now throws upon the subject. This part of Palestine had perhaps been more closely and generally studied than any other, thanks especially to the late Dr. Barclay of the American Mission. And yet if the Committee of the Palestine Exploration Fund needed a proof of the necessity for its exertions, it would be supplied by a comparison of their survey with the aforesaid map of the Environs of Jerusalem.

Below Kulonieh, the central Wady of the Nahr Rubin basin, runs as Wady es Surar southward and westward, in a deep and winding valley to Setûf, Kh. el Loz, and 'Akûr, receiving on its right bank between Loz and 'Akûr, the Wady esh Shemarin from Kustul and Soba; and on its left bank opposite 'Akûr, the Wady es Sikkeh, the recipient of the noted Wady el Werd, and Wady Bittir, which drain the eastern and part of the southern margin of the Rubin basin from Lifta and Jerusalem, to Bethlehem, el Khudr, and Kefr Sôm. West of Kefr Sôm, another wady descends to the left bank of Wady Surar now called Ismain, at Deir esh Sheik (alt. 1,595 feet). The Wady Ismain continues westward at the foot of Deir el Hawa (alt. 2,090 feet) to 'Artûf (alt. 910 feet), where it receives the Wady en Najîl direct from the southern edge of the basin, near Beit Nettif. This northward course of the Najîl will be mentioned hereafter in connection with other features, which serve to define the base of the mountains, and separate them from the lowland hills of the Shephelah. A tributary of the Najîl skirts the southern edge from the south of Kefr Sôm; and a parallel tributary on the north joins the Najîl from the same direction, passing

Beil Atâb and Jerash, Deir Aban and Deir el Hawa also send branches to the Najîl.

At Artûf also falls into the right bank of Wady el Surar, the important Wady el Ghurab, which drains the northern waterparting of the Rubin basin that extends from Beit Surik westward through Kuriet el Enab, Saris, and Beit Masir. The Ghurab washes Beit Nakuba, el Ammûr, Kesla, Eshua, and Artûf, and receives small branches from Soba and Kuryet el Enab in its upper part, and from Surah near its outfall. Between Eshua and Surah a branch of Wady el Ghurab comes direct south from the northern waterparting, and forms a part of the long depression that separates the mountains of Judah from the lowland hills.

Between the deep and precipitous valleys of the Ghurab and the Surar, is a rugged mountain which appears to correspond with the Mount Seir of Joshua. On the north of the Ghurab is perhaps the Mount Ephron of Joshua. It is the range which forms the waterparting of the Auja and Rubin basins, between the heads of Wady Ghurab and Wady Amir and those of Wady Selman and Wady Aly.

Below the confluence of the Ghurab, the Wady es Surar crosses a plain in the midst of Samson's country and surrounded by many biblical sites. This plain is a prominent part of the depression which runs north and south between the foot of the mountains of Judah on the east and the lowland hills of the Philistine Shephelah on the west. Where the Wady es Surar enters the lowland hills, it has the remains of Ain Shems (the biblical Beth Shemesh) on a southern eminence; while Surah, the Zorah of Samson and the tribe of Dan, looks down upon it from the north. The wady passes through the hills for about nine miles in a north-westerly direction, and skirts the southern edge of the plain of Akir (Ekron), which has a low ridge on the west, with a gap between el Mughar (alt. 236 feet) and Katrah (alt. 195 feet). Through this gap the Wady es Surar passes on its way to Yebnah (Jabneel) and the sea; near which it takes the name Nahr Rubin, from a shrine dedicated to the Patriarch Reuben.

The north-west waterparting of the Nahr Rubin basin runs along the summit of the hills on the north of Wady Surar, from Beit Mahsîr, to Abu Shusheh, and beyond Ramleh. From this slope many affluents join Wady es Surar. 1. From the extensive ruins of Râfât, and Surik near Surah, comes a wady which may be identified with Samson's Valley of Sorek, although the Wady Surar itself has hitherto been adopted. Samson's En Hak-kore, is also placed at Ain ek Kharyeh or Ayun el Kharjeh, and his Ramath Lehi at Kh. Ism-Allah.* Mahaneh Dan or the Camp of Dan, is probably at Khurbet Kila or the Ruined Fort. These are on the Wady el Khayeh, which becomes Wady el Baght towards its junction with Wady Surar. 2. Other wadys fall in below Khuldeh. 3. The slope around the plain of Akir (Ekron), is drained by many channels which all unite near Ekron, and join the Surar at the gâp below el Mughar. These include the villages of Sidun, Nâ'aneh, el Mansurah, and Akir. The Wady Deirân and the Wady Ayun Do, drain the hilly tract between Ramleh and Yebnah, and join the Nahr Rubin about two miles from the sea.

On the south or left bank, the Wady es Surar receives the Wady en Nahir, on the west of Ain Shems. It comes from the southern waterparting at Beit Nettif (alt. 1,517 feet), and passes the remains of Jarmuth and Zanoah. West of the Wady en Nahir, the southern edge of the basin is drained by the Wady el Menâkh which descends from el Bureij (alt. 830 feet), Mughullis, Jilia, and Kezâzeh to enter the Surar near Shahmeh (alt. 186 feet). Small drains also join the left bank at Katrah, Beshit, and Yebnah. These descend from a group of hills which culminate in a height of 305 feet, and correspond with the Mount Baalah of Joshua xv, 11, on the border of the tribe of Judah, which "went out unto Jabneel (Yabneh)."

* Conder's "Tent Life," i, 276-7.

THE BASIN OF NAHR SUKEREIR.

The delineation of this basin has undergone great changes from the Palestine Exploration Survey. Its northern border is coterminous with the Nahr Rubin basin. The eastern side, beginning on the south of el Khudr,* divides the heads of Wady Musurr on the west from the springs of Wady Urtas, which descend through Wady Derajeh to the Dead Sea. The eastern side continues southward along the edge of the Dead Sea basin, through Ras esh Sherifeh (alt. 3,258 feet) and along the road to Hebron, until the waterparting bends to the westward towards the village of Sâfa, whence it goes south again to rejoin the highway to Hebron, near Khurbet Jedur; then it passes Beit Ummar, Beit Sûr, and Hulhul, where this basin ceases to be connected with that of the Dead Sea, and follows the basin of Wady Guzzeh up to the south of Dura. Here and westward to Kh. Ejjis er Raz, the southern limits of this Sukereir basin divide it from the basin of Wady el Hesy, which falls into the sea on the north of Gaza. From Kh. Ejjis er Raz, the waterparting follows a line of hills up to Kh. Yasin, near Esdûd (Ashdod), which divides the Sukereir from the small basin of Wady el Bireh, a coast basin like that of el Falik, on the north of Jaffa. The basin of Wady Guzzeh is continued from Dura, along the eastern and southern sides of el Hesy.

The drainage of Sukereir basin is divided between three main branches.

1. The northern branch is Wady es Sunt, which takes the whole of the eastern drainage from el Khudr on the north to Hulhul on the south, and that was known before the present survey. The main channel of Wady Sunt begins with Wady Musurr, which rises on the west of el Khudr (alt. 2,832 feet). It takes the drainage of the north-east angle of the basin from Kh. Umm el Kulah on the north to Kh. Beit Skaria on the south. At Jeba it receives a feeder

* The Convent of Saint George, three miles west of Bethlehem.

from the northern margin at Kh. el'Ahbar and Wady Fukîn, and runs westward as Wady el Jindy to Kh. Shuweikeh (Shocoh). Here it receives a tributary from the northern margin of the basin, on the south of Kefr Sôm, which follows the margin westward as far as Beit Nettif. At Kh. Shuweikeh also is the junction of Wady es Sûr, which rises at Kh. Beit Nasif near Terkûmieh and runs almost due north, in a line prolonged by the plain of es Sunt, the Wady en Najil and the Plain of Surar, the Wady Ali, and the Plain of Beit Nuba to the foot of the Beth Horon Pass. This remarkable depression is about thirty miles in length, and serves to divide the mountains of Judah from the lowland hills of the Shephelah. The Wady es Sûr is the recipient of a succession of valleys which have their heads along the eastern margin of the basin, between a point west of Safa and Beit Sûr near Hulhul. The cave of Adullam according to Mons. Ganneau is at Aid el Ma in the Wady es Sûr.

Below Kh. Shuweikeh, the Wady es Sunt enters a gorge through which it zigzags on its way to the Plain of Philistia which it reaches at the foot of Tell es Safi, the Blanche garde of the Crusaders, and perhaps the Libnah of the Bible. The altitude of Tell es Safi is 695 feet, and the wady at its foot is 395 feet above the sea. In its passage through the gorge it receives a valley from the south, traversed by the Roman road from Beit Jibrin. Near its entrance into the plain it receives another affluent from Kudna (alt. 810 feet), Rana (alt. 660 feet), Dhikerin (alt. 680 feet) and Deir es Dhiban (alt. 665 feet). It crosses the rolling plain in a north-westerly direction to Jisr Esdûd or the bridge of Ashdod, where it joins the outlet of the central and southern parts of the basin.

2. The central branch is best known as Wady el Afranj (Feranj of former maps). This branch drains the eastern waterparting, where it abuts on that of Wady Guzzeh, about the heads of Wady el Khulil, around Hebron. It rises from two small feeders skirting the border on the east of Dura, and

it receives five tributaries which rise along the eastern border to the northward of the main sources. The northernmost is the chief of these. It has its most northern source in the Wady Kaideh which rises close to the Hebron road on the west of Hulhul. But the most easterly source of this tributary is about half a mile further south, on the east side of the Siret el Bellâ'a (alt. 3,370 feet), a prominent height on the east of the Hebron road. Lower down this tributary skirts Terkûmieh and joins Wady el Afranj on the north of Idhna.

This branch has been supposed hitherto, to belong to a drainage system, quite distinct from Wady es Sunt. On leaving the hills at Zeita, instead of running as it does to the north-west to join the Sunt at Ashdod, the Afranj was reputed to cross the plain towards the south-west, and to join the Wady el Hesy at Simsim. See map of Holy Land in Dr. William Smith's Ancient Atlas ; also Lieutenant Van de Velde's map ; and Dr. Robinson's " Phys. Geog.," 107.

3. The southern branch of the Sukereir basin, rises from south-eastern angle at Dura, and receives tributaries from the margin of the Wady el Afranj north-westward to a point on the west of Beit Jibrin. It is the Wady ed Dawaimeh of former maps, and was so called from a village of that name on the main channel. On the east of this place it receives the drainage from between Dura and Idhna. At el Kübeibeh it receives the drainage from the margin of the Wady el Afranj, between Idhna and Beit Jibrin. From Kubeibeh, it flows north-westward to 'Arâk el Menshiyeh, where it receives a branch rising on the west of Dawaimeh, and taking the drainage of the southern edge of the basin from its source to Kh. Ajlan (biblical Eglon). Another affluent rises at Kh. Ajlan, and taking the drainage of the western border up to Kh. Mamin, joins the main channel at Keratiya. At Keratiya the main southern branch is called Wady el Ghueit, and runs on north-westward to join the el Afranj at Beit Duras, from whence the junction with the northern channel is made at Jisr Esdud. The passage to the sea then proceeds northward for three miles, when it finally turns to the west and

E

reaches the coast at Neby Yunis, a shrine of the Prophet Jonah.

Of course this southern branch of the Sukereir basin was supposed to follow the central one to the basin of Wady el Hesy; which indeed has been entirely deprived of its reputed channels, and has been furnished with a newly discovered series by the present survey. Such was the recent state of the geography of an important part of Philistia and Judæa.

The Basin of Wady el Hesy.

The outfall of Wady el Hesy lies between the ancient cities of Gaza and Ashkelon. The northern limit of its basin runs with the Sukereir basin, and beginning at the sea, and the sandhills at the mouth of the river, proceeds north-eastward between Berberah and Nalia to Kh. Erzeh, Khurbet Samy and Kh. 'Ejjis er Ras (alt. 331 feet). [These places mark the division between Wady Kemas in the basin of el Hesy, and Wady Bireh. The latter belongs to a small but distinct basin, including the villages of Nalin, Askalan (Ashkelon), el Mejdel, Hamameh, and Julis. Its outlet is at Tellul el Ferani (alt. 50 feet) where it appears to be lost in the sands.]

From Kh. 'Ejjis er Ras (alt. 331 feet) the northern waterparting turns to the south-east and south, through Neby Ham to Khurbet Melita (alt. 336 feet). Here it strikes south-eastward again to Kh. Ajlân (Eglon), and more easterly along a succession of ranges by Tell Jabîs (alt. 450 feet), Muntaret el Kaneiterah, and by a long sweep to Khurbet er Resum (alt. 1,090 feet). From this point it bends round the heads of the Valley of Dawâimeh, and then makes another bend around the valleys on the west of Dura; finally reaching Ras el Biain (alt. 2,950 feet).

At Ras el Biain, the eastern side of the waterparting begins, dividing the Hesy from the Ghuzzeh (Gaza) basin. It passes south to Khurbet Kharsa (alt. 2,857 feet), Kh. Sirreh (alt. 2,746 feet), Ras Sirreh (alt. 2,601 feet), and Merj

Dômeh, to a point about one mile north of edh Dhâheriyeh; whence it proceeds westward to Sheik abu Kharrubeh (alt. 2,070 feet), then south-west to Ras en Nukb (alt. 2,023 feet), and on to Tell Khuweilfeh, the southernmost point of the basin.

Here the waterparting strikes north-westward by the track of Kanan es Seru to Kh. Umm Ameidat (alt. 560 feet), and Tell Abu Dilâkh (alt. 470 feet); still continuing along the Kanan es Seru to the heights of Saleh Burber; whence the waterparting strikes southward to Kh. Zuheilîkah (alt. 450 feet) and Kh. el Jindy; then north-west to Kh. Hurâb Dîâb (alt. 490 feet); and westward, near Khurbet Sihân and Kh. Mansûrah to the gardens of Gaza, and to the sea.

From Tell Khuweilfeh westward, the waterparting divides the affluents of Wady el Hesy from the Wady esh Sherîah, which runs to Wady Ghuzzeh, and forms the northern boundary of the pastoral nomadic tribes of the 'Azâzimeh and the Teiâha. West of the 'Azâzimeh are the Terâbîn, whose territory reaches from Gaza to Suez.

The Watercourses of the Basin of Wady el Hesy.

The main channel has its origin at the north-eastern extremity of the basin. Six wadys rise along the eastern waterparting between Ras Biain (alt. 2,950 feet) and Ras Sirreh (alt. 2,061 feet). These unite at different points on the east of Kh. 'Aitûn, and flow north-westward, receiving an affluent from Beit Auwa (alt. 1,495 feet) on the northern edge of the basin. The next of importance falls in on the left bank, from a wady which has its sources along the eastern waterparting, between Ras Sirreh and Sheikh abu Kharrubeh (alt. 2,070 feet). The main channel hugs the northern edge of the basin, and has only short tributaries on that side, till it reaches the plain at Simsim (alt. 219 feet), where an outfall takes place from Umm Lakis (Lachish), el Huleikat, and Bureir.

At Deir Sineid, a longer branch falls in, which rises at

the northernmost part of the basin at Kh. 'Ejjis er Ras (alt. 331 feet), and takes the drainage of Kaukabah, Beit Tîmn, Ejjeh, Burberah, and Beit Jerjah.

From the southern waterparting, between Sk. abu Kharrubeh and Tell Khuweilfeh, the Wady edh Dhikah proceeds, and takes the name of Wady en Näs on the way to its junction at Kh. Surrâ. Another considerable branch rises near the same part, and skirts the waterparting up to Kh. el Mukeimin, when it turns to the north-west and passes Kh. abu Gheith (alt. 640 feet) on its way to the main wady at Tell el Hesy (alt. 340 feet). The modern name of Gheith, its position on the southern frontier of Philistia, midway between the ancient fortresses of Hebron and Gaza, and particularly its connection with "the way to Shaaraim" (Tell esh Sheriah) have caused the remains of Gheith to be regarded as the representative of the long lost, ancient Gath, one of the five cities of Philistia, and the birth-place of Goliath.

Another affluent from the south, rises at Kh. Umm Ameidât, waters the modern village of Huj, and joins the Hesy at Khurbet Jelameh. The next from the southern waterparting, rises at Kh. Zuheilikah (alt. 450 feet) runs nearly up to Gaza where it receives a tributary from the south, and bending northwards, reaches the Hesy at Deir Sineid.

THE BASIN OF WADY GHUZZEH.

The Palestine Exploration Survey only includes the northern part of this basin. Where the southern limit is to be drawn, it would be rash to affirm with any pretension to certainty, after the proofs supplied by the Survey of the unreliable character of existing maps dependent on route surveys in respect to questions of this precise character. The present southern limit of the survey corresponds with the division between Philistia and the Hill Country of Judah on the one hand, and the exceedingly interesting but unsurveyed and therefore very imperfectly known region including the

Negeb or "South Country" of Scripture on the other. How attractive that part of the Holy Land is to the student of the Bible, is amply attested by that admirable work, which the Rev. Edward Wilton, M.A., Oxon., has devoted to it.* That book alone yields abundant evidence of the desirability of extending the survey work of the Palestine Exploration Fund to the southward. No doubt it is an undertaking which must be approached with ample precautions, but there is sufficient evidence to warrant the belief that a good understanding for the purpose, might be entered into with the chiefs of the tribes, whose goodwill it would be necessary to secure. They are able to appreciate reverence for the Sacred Writings. It would be possible to explain to them the desire of believers to recover such a knowledge of the sites of the holy places in the "South Country," as the Survey now supplies further north. Their objections and apprehensions may be ascertained and provided for. A conviction that the welfare and prosperity of these pastoral people would be studied and promoted by the supporters of the Survey, might be justly impressed upon them. Such are some of the notions that seem to encourage the confident expectation that the Survey of the "South Country" of the Holy Land, will soon be taken up with the same earnestness and ability, which has already brought so large a part of the Fund's work to a most successful issue.

The northern limit of the Wady Ghuzzeh (Gaza) basin, runs with that of el Hesy from the coast eastward to Kuweilfeh, and thence north-eastward to Ras Biain; beyond which it runs in the same general direction nearly to Hulhul, on the north of Hebron, having in this part the basin of Nahr Sukereir on the west. Here it becomes attached to the basin of the Dead Sea, and runs southward along its margin through Beni Naim, Tell ez Zif, el Kurmul, Tell Main, Khurbet el Kureitein, Khurbet Beiyud, and

* The Negeb or "South Country" of Scripture. By the Rev. Edward Wilton, M.A. Oxon., Macmillan & Co., 1863.

Tell Arad, where the map terminates, leaving the rest of the waterparting really unknown. Every one of the names along this eastern margin of the basin are sites of biblical interest and correspond to Janum, Ziph, the Carmel of Caleb, Maon, Kerioth Hezron, Beth-Lebayoth, and the Canaanite capital Arad.

The southern limit of the Palestine Exploration Survey, runs from the Dead Sea by Wady Seiyal, corresponding to the Wady Hafaf of Wilton, thence by Wady el Kureitein to Khurbet el Milh, the Moladah of Scripture, thence by the Wady es Seba to Bir es Seba, the biblical Beersheba, and the junction of Wady esh Sherîah, whence the Wady Ghuzzeh runs on to the sea near Gaza.

The Watercourses of the Basin of Wady Ghuzzeh.

The head of the Wady el Khulil is the origin of the principal channel of the northern part of the basin. It commences in three wadys which unite at Hebron, and in others further north towards Khurbet Beit Anûn, the biblical Beth Anoth (alt. 3,085 feet), which unite on the north-west of Hebron, and contribute to the Wady el 'Aawir, which joins the Hebron Wady, a few miles south of the city. The Wady el Khulil zigzags from this confluence south-westward to Rujm ed Deir (alt. 2,612 feet) near Yutta (alt. 3,747 feet), the scriptural Juttah, and, according to Reland, the homely retreat of Zacharias and Elisabeth and the birth-place of their son John the Baptist. The bottom of the Wady el Khulil is thus 1,135 feet lower than Yutta, near which it receives the Wady Kilkis and also the Wady ed Dilbeh on the right bank. Both come from Khurbet Kanan, on the waterparting south-west of Hebron, and both are partly skirted by the high road which further south impinges on the Wady el Khulil, where the channel encircles the remains of Khurbet Rabud. Here it receives an affluent on the left bank. The Wady Khulil continues down the deep valley in a winding course to the foot of a spur surmounted by a track from the village of edh Dhaheriyeh (Debir of Caleb) on the western hill-side (alt. 2,150 fcet). This is the first village in Palestine, by the road from Sinai through Beersheba and Hebron. At the foot of the spur, the Wady Deir el Loz falls in on the left bank. The wady continues to wind about in its south-westerly descent, receiving the outfall of a group of wadys which come from the waterparting between Ras Sirreh and Ras en Nukb (alt. 2,023 feet), and passing Khurbet Tât Reit, skirt the north-western side of the Dhaheriyeh ridge. Afterwards it proceeds on a straighter course, and is joined by Wady Itmy from the northern edge of the basin at Kh. Khuweilfeh (Robinson's " Bib. Res." i, 207); finally reaching the Wady es Seba at Tell es Seba (alt. 950 feet) on the east of Beersheba (alt. 788 feet). This confluence is also joined by another wady from the north, which skirts the Khashm el Buteiyir.

The interval between the basins of el Hesy and the Dead Sea, widens between Dura on the west and Tell es Zif on the east, and gives rise to another system of watercourses, on the east of Wady el Khulil. It rises on the eastern water-parting, on the north of Tell ez Zif (alt. 2,882 feet) the biblical Ziph "immortalized by its connexion with David." It skirts the waterparting to el Kurmul (alt. 2,067 feet) the Carmel of Saul,* David,† and Uzziah,‡ and, appearing again in the history of the Crusades. Here it turns to the west, receiving a tributary from Yutta (alt. 3,747 feet), and doubling upon itself, bends south to es Semua (alt. 2,407 feet), the Eshtemoa of David's exile, near which it receives tributaries from the eastern waterparting about Maon (alt. 2,887 feet), and Bîr el Edd (alt. 3,000 feet). Pursuing its south-westerly course as the Wady el Khan, it reaches Zanuta, the biblical Zanoah, where it receives a branch from Kanan el Aseif (alt. 3,002 feet) and Râfât (alt. 2,312 feet). It passes Kh. Attir, the biblical Jattir (alt. 2,040 feet); receives the Wady el Habûr,

---

* 1 Sam. xv. 12.
† 1 Sam. xxv. 2, 5, 7, 40; xxviii. 3; 1 Chron. iii. 1.
‡ 2 Chron. xxvi. 10.

the Wady el Ghurra, and the Wady Saweh, and joins Wady es Seba near Tell es Seba.

Above this junction, the Wady es Seba may be traced to its origin on the eastern waterparting about Khurbet el Kureitein; from whence it runs southward as Wady el Kureitein to Khurbet el Milh, the site of Moladah (alt. 1,210 feet), where it bends abruptly to the west on its way to Beersheba.

From the confluence of the Wady el Khulil, the Wady es Seba proceeds by the west through the pastures of the 'Azâzimeh and Terâbin Arabs, to its junction with a great wady from the south, which will be an attractive feature in the extension of the Survey. At this point the Wady es Seba changes its name to Wady Guzzeh, and runs north-westward to its confluence with Wady esh Sheriah, and its further passage to the sea on the south-west of Gaza.

The Wady esh Sheriah drains the northern margin of the basin of Wady Ghuzzeh, from the sea to Tell Khuweilfeh. On the east it has the Wady Itmy and another affluent of Wady el Khulil, being separated from the latter by Tuweiyil abu Jerwal and Khashm el Buteiyir, names which are applied to a spur which stretches southward from the range of the northern waterparting to the Sahel Umm Butein or plain of Beersheba. The southern boundary of Wady Sheriah is the summit of a broad down, which undulates between the Sheriah and the Seba.

Two main branches divide the basin of Wady esh Sheriah and meet at Khurbet 'Erk. The northern branch is the recipient at Kh. Umm el Bakr, of several wadys which rise along the north-eastern edge of the basin, from Kh. Umm Dabkal round by the east to Tuweiyil abu Jerwal (alt. 1,500 feet). The Wady Sheriah passes westward from Kh. Umm el Bakr to Tell esh Sheriah (alt. 400 feet), an ancient site identified with Shaaraim, the way to which place is connected with the long-lost city of Gath, in the account of the flight of the Philistines, after Goliath was slain by David, 1 Sam. xvii. 52. Between Tell esh Sheriah and the junction with the

Ghuzzeh seven wadys join the Sheriah from the northern edge of the basin.

The southern branch collects all the south-eastern affluents of the Sheriah and carries their waters to Khurbet Erk. Monsieur V. Guerin has contributed some additional sites to this part of the survey, which the surveyors left incomplete. But it is difficult to make his routes fit with the Palestine Exploration Map, and his own map like the best of others relating to this part, only serves to repeat the proof of the imperfect results of route surveys.

At the junction of Wady Sheriah with Wady Ghuzzeh, the survey places the Khurbet el Kutshan, which appears to be Arabic for the "Ruins of the Horse Village." This at once suggests its identification with the biblical Hazor Susah or Susim, which has the same meaning. The site has been long looked for in this locality. It has been supposed to be connected with the trade in horses with Egypt in Solomon's time; but its association with this neighbourhood is more fitly explained by the fact that horsebreeding is a prominent pursuit in the pastures about Gaza. The Henâdy Arabs of this part are famous for it.

On the south of Wady Ghuzzeh and near the sea, the Survey places the remains of Deir el Belah, which Mons. V. Guerin identifies with the Crusaders' fortress of Darum. The Rev. E. Wilson applies his cogent reasoning to connect it with the Bizjoth-Jah-Baalah of Joshua xv, the Balah of Joshua xix, and the Bilhah of 1 Chronicles iv. It is undoubtedly the site of a sacred fane of high antiquity, suitable for the worship of Baal, and still more to give expression to the contempt of the Almighty (Bizjoth-Jah) for that idolatry.

Here ends the present examination of the Survey within the Mediterranean watershed. The remainder includes the western slope of the Jordan and Dead Sea Basin. It will be subjected to the same analysis as the former part; and the comparison of the knowledge acquired by the Survey with earlier work will be not less instructive and interesting.

Part II.

THE WESTERN WATERSHED OF THE JORDAN AND DEAD SEA BASIN.

THE BASIN OF WADY ET TEIM AND NAHR EL HASBANY.

The head of the Jordan basin lies about twenty-four miles beyond the present limits of the Palestine Exploration Survey, among the sources of the Wady et Teim, and surrounded by the villages of Medukhah, Bekka, and 'Ain el Arab. Near it passes the high-road from Beirut on the Mediterranean, over Mount Lebanon to Damascus by way of the Wady el Kûrn. The wady runs to the Barada river, and divides the northern flanks of Mount Hermon from the southern extremity of Anti Lebanon.

The Wady et Teim drains the western slope of Mount Hermon; and, as the Nahr el Hasbany, it comes into the Palestine Exploration Survey. It has on its right bank the Jebel ed Dahar, a narrow ridge on a portion of the waterparting between the Jordan basin and the Kasimîyeh. South of Dibbîn, the plain of Merj 'Ayûn lies between the Kasimîyeh-Jordan waterparting and the southern prolongation of Jebel-ed-Dahar. The waterparting skirts the western side of the Merj, and is continued along the southern prolongation of the Kasimîyeh, as described in the notice of that basin. The Kasimîyeh rises more than forty miles further north than the Wady ed Teim, on the flanks of the highest summit of Mount Lebanon; from whence it descends along the eastern base of that mountain, till it passes the Crusaders' fortress of Belfort now Kulat esh Shukif. There it turns abruptly westward to the sea, and falling within the limits of the Survey, it came under notice at the commencement of this investigation.

The Nahr Hasbany and the Wady et Teim are without a record in ancient geography. They are discoveries made since the Palestine Association began the work which has been taken up by the Fund. The study of watersheds and basins commenced since accurate surveying supplied the requisite information, just as anatomy originated with the precise examination of the human frame. Formerly nothing was known of the Jordan further north than the fountains which supply its perennial waters on the south of Mount Hermon. The chief of these gushes out of the western side of Tell el Kâdy,* or the Judge's Mound (alt. 505 feet), and is one of the largest in the world, while another springs from the top of the same Tell directly above, and forms a distinct and considerable stream running to the southwest, and driving two mills, before it joins the other river. The ruins on the Tell are the remains of Dan, the northern counterpart of southern Beersheba, the foundation of which is recorded in the Books of Joshua and Judges. Its name is still retained by its fountains and stream, the 'Ain and Nahr el Leddân. The change from Dan to el Leddân is plainly traced by Dr. Smith in a note quoted by Dr. Robinson ("Bib. Res.," iii, 392), and Dr. Wilson remarked that Kady and Dan are respectively Arabic and Hebrew for a judge ("Lands of the Bible," ii, 172).

The next fountain in importance springs up at Banias (alt. 1,080 feet) in a nook of the mountain at the inner or northeastern angle of the terrace, on which are the remains of this ancient place. The stream is called Nahr Banias and joins the Leddân in the plain. Not far below the confluence, a third affluent adds to the bulk of the stream; it is the Nahr Hasbany which comes from the Wady et Teim and the northern extremity of the basin. Near its outfall (alt. 140 feet) the Nahr Hasbany receives (1) the Nahr Bareighit, which has its source in the Merj 'Ayûn at 'Ain ed Derdêrah near the Kasimîyeh, and a few miles above its great bend. The Bareighit becomes nearly dry in autumn. (2) The

* An artificial looking mound of limestone rock, flat topped, eighty feet high, and half a mile in diameter, its western side covered with a thicket of reeds, oaks, and oleanders. Tristram's "Land of Israel," 580. See also Monsieur Guerin, "Galilée," iii, 338.

outlet of fountains rising along the eastern foot of the hills, above el Khâlisah and en Na'ameh. Dr. Robinson considers the Banias stream to be twice the bulk of the Hasbany, while the Leddân is twice that of the Banias. Dr. Wilson found the Hasbany to be seven yards broad and about two feet deep. The Leddân measured ten yards wide and two feet deep. At Banias, Dr. Wilson only remarked that the spring appeared to be "about as copious as that of Dan." According to the Memoirs of the Survey, the Nahr Banias is the principal source, and the fountains at Tell el Kady are mentioned as one in "Memoirs," 17A, but not in p. 24.

THE HULEH PLAIN, MARSH, AND LAKE.

The confluence of the perennial streams which unite to form the Jordan of the Bible and History,* takes place on the Huleh Plain, described on p. 144; and the river enters the marsh and lake of el Huleh, at es Salihîyeh. Here are the Waters of Merom (Josh. xi, 5, 7); and Lake Semechonitis of Josephus. Both lake and marsh have been examined by Dr. Tristram who nearly lost his life in the marsh; and by Mr. Macgregor who penetrated them at great risk in the "Rob Roy" canoe. Tristram's "Land of Israel," 585-590. Macgregor's "Rob Roy on the Jordan."

The principal wady entering the marsh from the west drains a recess of the waterparting, which affords space for the plateau of Kades, the famous Kedesh of Naphtali. On the north of this plateau is one still more elevated, and having no outlet for its waters. It contains the villages of Meis, and its waters form lagoons in the rainy season. Robinson's "Bib. Res.," iii, 369. From the south-west angle of Meis plateau, the Kades plateau flanks the most southerly part of the Kasimîyeh basin between Meis and Aitherun. It is the recipient of several branches before it descends from the hills as Wady Arus. Another wady from the hills unites with the abundant waters of 'Ain el Mellahah,

---

* As distinguished from its extension northwards throughout Wady et Teim.

HULEH PLAIN, ETC. WADY EL HINDAJ. 65

which enters the lake through the southern edge of the marsh. This wady does not impinge on the Mediterranean waterparting, but is divided from it by the next; and as a rule only those wadys will be noticed separately hereafter, which are in contact with the Mediterranean basins.

THE BASIN OF WADY EL HINDAJ.

The western side of Lake Huleh receives one wady from the Mediterranean parting, and two which are divided from it. The first is known in its lower part as the Wady el Hindaj. On the north the waterparting commences on the edge of the lake at Tell Abalis, and touches on the plateau of Kedes, passing the southern extremity of the Kasimîyeh basin on its way to Marûn er Ras (alt. 3,035 feet). Turning southward, the basin of the Hindaj, runs with the Ezzîyeh basin between Marûn er Ras and Sasa. From Sasa the boundary of the Hindaj ascends to the top of Jebel Jurmuk (alt. 3,934 feet), the culminating summit of Galilee. Here the Hindaj adjoins the Mediterranean basin of Wady el Kûrn. After descending the northeastern slope of Jebel Jurmuk, the boundary goes towards Ras el Ahmar, and reaches Lake Huleh at et Teleil. Between Jebel Jurmuk and Ras el Ahmar, the Hindaj basin runs with the northernmost part of the Wady Amud or Safed basin, which falls into the Sea of Galilee. Eastward of Ras el Ahmar, it is bounded by the minor basin of Wady Shebabik. This basin is more fully noticed in pp. 185 to 188.

The basin of Wady el Hindaj includes the village of el Jish, the ancient Giscala (alt. 2,370 feet), and also those of Farah (alt. 2,160 feet), Salhah, 'Alma, and Deishun. The biblical sites of Edrei and Hazor are reputed to be in this basin, at Hadireh and el Khureibeh.

MINOR BASINS—WADY SHEBABIK, &C.

The Wady Shebabik is the most northerly of a series of minor basins which are divided from the Mediterranean Slope by the basin of Wady Amud or Safed. It is followed by Wady Musheirefeh, properly Loziyeh, and both fall into Lake Huleh. At Kh. Benit, between the heads of Shubabik and Loziyeh, is a commanding view over the Huleh basin. Robinson, " Bib. Res.," ii, 434. The Wady Loziyeh, properly Musheirefeh falls into the Jordan on the south of Jisr Benât Yakûb, or the Bridge of Jacob's daughters. Besides it, this part of the Jordan only receives a few short wadys. The remainder of this secondary series falls into the Sea of Galilee. The first is Wady Zuhtuk, which runs nearly parallel with the Jordan, and joins the lake near it. The next, with a trifling intervention rises in Jebel Kanan, on the east of Safed, and enters the sea at Tell Hûm. The Wady Jamus has its mouth at 'Ain Tabghah.

The Basin of Wady 'Amúd, Safed.

This is an important basin, containing the noted town of Safed. It comes into actual contact with the Mediterranean slope on the margin of Wady el Kûrn, but it is only divided from the Kasimîyeh on the north by the Hindaj, and from the N'amein and the Mukutt'a basins on the south, by the Rubudiyeh basin. It enters the sea through the plain of Ghuweir or Gennesaret, at Tell el Henûd. See p. 186.

The Basin of Wady Rubudiyeh.

Its mouth is not far south of the preceding wady. The head of the basin is spread between Jebel el Arus (alt. 3,520 feet) and Jebel Abhariyeh, where it is divided from the Wady el Kûrn on the north, and on the east from the Plain of Rameh, watered by the principal branch of the Wady Halzûn in the Naiman basin, which empties itself near Acre. On the east it has the Safed basin. At Ailbûn, on its southern border, it approaches the Plain of Buttauf, in the north-eastern part of the Mukutt'a basin. From Ailbûn to the Sea of Galilee it is bordered by the Wady el Hamâm.

The Basin of Wady el Hamâm.

The Mediterranean waterparting here approaches closely to the Sea of Galilee and confines this basin within a short extent. It is, however, of much interest, for it rises near Hattin, where the Crusaders suffered a decisive defeat; and includes the site of Irbid, and also the biblical Magdala, now el Mejdel, where it falls into the sea.

A series of minor basins skirt the rest of the coast of the Sea of Galilee between Wady el Hamâm and Wady Fejjas, which falls into the Jordan at the Jisr es Sidd. The only one at all notable is the Wady el Amis next to Wady el Hamâm. For from Tiberias to Jisr es Sidd the basin of Wady Fejjas skirts the sea closely, and at a height which attains to 1,650 feet above its depressed surface, leaving only a steep and narrow margin furrowed by precipitous channels towards the shore. The level of the Sea of Galilee as determined by the Survey is 682·5 feet below the level of the Mediterranean.

THE BASIN OF WADY FEJJAS.

The waterparting of the Jordan is thrust inward at the head of this basin, which is coterminous with the most easterly extension of the Mukutt'a at the Plain of Toron. It has however some length, in consequence of its oblique direction from north-west to south-east. On the south it is bounded by the more considerable basin of Wady Bireh, except towards the outfall into the Jordan at the Jisr es Sidd, where a few short secondary channels are interposed along the right bank of the Jordan between the outfalls of this wady and Wady Bireh. Among them is the Wady umm Walhan with a permanent stream falling from a height of 2,000 feet in a short distance. It falls into the Jordan on its right bank, about a mile above the junction of the Yarmuk on its left bank, where the depression of the valley below the level of the Mediterranean is 835 feet. About two miles lower down, the Jordan is crossed by the Jisr el Mujâmia, on the road to the Yarmuk and Um Keis (Gadara).

THE BASIN OF WADY EL BIREH.

The heads of this basin extend along the margin of the Mukutt'a between the villages of esh Shejerah and

Nain. On the south-west it is bounded by the heads of Nahr Jalûd, until the secondary basin of Wady Yebla or 'Esh-sheh ('Osheh formerly) intervenes, along with three others of no magnitude.

The principal channel of this basin rises at its northern extremity and flows at the eastern base of Mount Tabor, receiving on the south-west of the mountain another branch which rises on the east of Nazareth, and descending between Iksal and Deburieh, the Chesulloth and Dabberath of Scripture, passes the south of Mount Tabor to join the northern branch. There are many tributaries on both sides of the main stream, and one of them comes from the biblical Endor. The river passes from the hills into the Ghor or Valley of Jordan, by a fine gorge which has Kaukab el Hawa "the Star in the Air," on the southern summit. It is the remains of the Crusaders' Castle of Belvoir, and the ruins are occupied by a miserable peasantry.

THE BASIN OF NAHR JALÛD.

The permanent stream rises at 'Ain Jalûd and 'Ain el Meiyiteh, near the village of Zerin, the ancient Jezreel. It waters the noted Valley of Jezreel, and the village of Beisân, the site of biblical Bethshean, and the later Scythopolis. Below Beisân it crosses the Ghor to enter the Jordan through the ravine of ed Duwaimeh.

The head of the basin lies between Jebel Duhy (alt. 1,690 feet), and Jebel Fukû'a or Mount Gilboa, which at Sheikh Burkan is 1,698 feet. It skirts the edge of the plain of Esdraëlon through the villages of el 'Afuleh and Zerin. The plain is called also the Valley of Megiddo, and by the present inhabitants, Merj Ibn 'Amir. Esdraëlon is the well known Greek form of Jezreel, and the " Plain " which extends from Zerin westward, must be distinguished from the " Valley " which descends rapidly from it, eastward to the Jordan.

The descent of the valley is thus defined. Zerin is 402 feet above the sea. The 'Ain Meiyiteh at the foot of the village, is only 60 feet above the sea; the 'Ain Jalûd within two miles of Zerin, is 120 feet below the sea level. Beisân is on the edge of a broad terrace, which extends southward along the foot of the mountains for several miles at a height of 322 feet *below* the sea. The terrace has a steep descent to the Ghor or upper valley of the Jordan, which is here between 700 and 800 feet below the sea. The edge of the terrace above the Ghor, is traversed by the ancient road between Nablûs and Beisân. The River Jordan itself runs in a narrow trench through the Ghor, at a still lower depth, which does not appear to have been observed nearer than Jisr Mujâmia, minus 845 feet, and at the foot of the ancient road which leads from the Jordan south-westward to Wady Farrah. At this point the river is minus 1,080 feet, which would make it about 950 feet below the sea near Beisân.

From Sheikh Barkân to Beisân and the Jordan, the southern edge of the Nahr Jalûd basin is undistinguishable among an intricate network of irrigation works and neglected swamps, which extend from Beisân southward to Wady Shubash.

The wadys descending from Mount Gilboa (Jebel Fukua) to Nahr Jalûd appear to be mere seams in the side of the mountain and require no further notice.

The wadys from the northern edge of the basin are more remarkable. The head of the basin—including the south side of Jebel Duhy (alt. 1,690 feet) and the edge of the valley passing through el 'Afuleh to Zerin—is drained by the affluents of Wady el Hufiyir, which with another distinct wady from Jebel Duhy, joins the stream from 'Ain el Meiyiteh, before its junction with the waters of 'Ain Jalûd.

From the slopes east of Jebel Duhy and around the village of en Naûrah, the Wady es Sidr descends to 'Ain Tub'aûn, a spring which rises close on the left bank of the Nahr Jalûd, facing 'Ain Jalûd on the right bank. The Wady es Sidr does not however join the Nahr Jalûd, for it maintains an independent and parallel course as an aqueduct called Kanat es Sokny as far as the Khan el Ahmar, where it runs off northward of the village to the Wady el Khaneizir.

The discovery of 'Ain Tub'aûn is of historical interest. It

solves a passage in William of Tyre where Saladin is said to have encamped by a fountain called Tubania, at the foot of Mount Gilboa, near Jezreel; "circa fontem cui nomen Tubania," etc.* The same event is related by Boha-ed-din in his Life of Saladin, as having taken place at 'Ain el Jalût, or Ain Jalûd, which considering the proximity of the fountains, involves no real discrepancy. Mons. Guerin in an instructive notice of 'Ain Jalûd, draws attention to these passages and concludes that 'Ain Jalûd is meant by both of the historians.† Thanks to the Palestine Exploration Survey, it will now be seen that the old French chronicler was quite exact, and that the name which he records exists to this day.

The Kanat es Sokny continues to intercept all the drainage of the northern slope. It is sufficient to note that the Wady el Harriyeh drains the villages of Kumieh and Shutta.

THE BASIN OF WADY SHUBASH.

The head of this basin is in contact with the south-eastern extremity of the Mukutt'a at Jelkamus, and with the north-eastern end of the Nahr Mefjir basin between Tannîn and Ras Ibsik. One of its sources descends from Ras Ibsik (alt. 2,404 feet) and the secluded village of Raba. Another rises on the north of el Mughaîr. The wady descends from these elevated glens by a gorge, which terminates at the south-western extremity of the Beisân Terrace, where it appears to end in a continuing slope with the lower level of the Ghor. It also receives branches from the projecting hills on which the village of Khurbet Ka'aûn is situated, which although 213 feet below the Mediterranean, is still about 700 feet above the Ghor at its foot. This Ka'aûn is probably the Coabis of the Peutinger Tables, in this direction.

THE BASIN OF WADY KHASHMEH.

The basin of the Shubash is succeeded by the Wady Khashmeh, and its affluent the Wady Selman, which unite at the foot of the mountains, below the village of Berdeleh.

* "Hist. Belli Sacri," lib. xxii, cap. xxvi.　　† Guerin, "Samarie," i, 309.

The Khashmeh barely touches the Mediterranean water-parting at Ras Ibsik, being almost wholly intercepted by Wady Shubash on the north, and by the heads of Wady Mukhnawy on the south. The watercourse seems to come to an end in the Ghor, without reaching a collection of five fountains, amidst the ruins named el Fatûr, ed Deir, and Umm el Amdan, which unite in a single channel passing direct to the Jordan. There are wadys on either side of Khashmeh, but these only rise on the hill-side skirting the Ghor.

THE BASIN OF WADY EL MALEH.

Between Ras Ibsik (alt. 2,404 feet) and Ras el Akra (alt. 2,230 feet), this basin is in contact with the northern arm of the Mefjir Basin. On the south of Mount Akra, it is divided from the tributaries of the Merj el Ghuruk, which has no outfall to the sea.

The northern boundary or waterparting extends from the confluence with the Jordan westward to the southern flanks of Ras Ibsik where it turns to the south-west, over Ras el Akra to a point east of Judeideh.

The southern boundary of the basin starts from its contact with Merj el Ghûrûk, between Judeideh and Tubâs, reaches Tubâs (alt. 1,227 feet), runs south to ed Deir, then east to Ras Jadir (alt. 2,326 feet), and pursues the summit of this range to Kh. Umm el Kotn (alt. 342 feet); thence it passes eastward to Kh. Mofia (alt. 590 feet).

At Kh. Mofia, the eastern boundary commences, and runs north to Ras Umm Zokah (alt. 840 feet) continuing in the same direction to Tell Fass el Jemel, and onward until the range of hills bends round to the east, following the course of the Wady el Maleh up to the Jordan. Thus the basin forms an irregular triangle, with its faces towards the north, south-west and east, and its outfall at the north-east angle.

This basin is drained by three main branches, viz.—(1) the Wady el Maleh, (2) the Wady Helweh, and (3) the Wady ed Duba.

The Wady el Maleh has its sources at the western extremity

of the basin. Two wadys extending in the same line, descend from Ras Akra and Ras Jadir in opposite directions, and meet in a fine elevated plain on the north-west of Tubas (alt. 1,227 feet). Running on northward, this wady is met by another coming towards it from Ras Ibsik (alt. 2,404 feet), and both turn eastward and descend through a gap in the hills to meet at Teiasir (alt. 995 feet) which overlooks another fine plain stretching out eastward. The wady advances round this plain by the north-west and then by the south-east, taking the names of Wady Mukhnawy and Wady el Hirreh, and receiving a tributary from the south side of the plain at a point between Kh. el Akâbeh (alt. 732 feet) and Burj el Maleh (alt. 718 feet). From this confluence the wady takes the name of Wady el Maleh, and advances eastward to the end of the range of Ras er Raby on which stands the Burj or ruined castle of Maleh. It receives on its way affluents from Ras er Raby on the north, and Ras Jadir (alt. 2,326 feet) on the south, the former washing the south-western face of the castle hill, and the latter passing the village of Kh. Yerzeh (alt. 950 feet). On the south-east of Maleh Castle, the wady receives another tributary from the range of Ras Jadir, the southern margin of the basin; and bends round to the north-east. At the eastern foot of the castle range, it receives a tributary which washes the north-eastern face of the castle hill, and thence it proceeds eastward to 'Ain Maleh and 'Ayûn el Asâwir, where it receives the Wady Helweh from the south.

The Wady Helweh rises along the south-western edge of the basin from Kh. Mofin (alt. 590 feet) to a point on the north of Kh. Umm esh Sheibik. These wadys meet at the foot of Kh. Umm el Hosr and run on northward to the confluence with Wady Kaû Abu Deiyeh from the eastern margin, at the foot of Kh. Umm el Ikba (alt. 276 feet), where two tributaries from the west also fall in. The Wady Helweh continues northward to 'Ain Helweh and to the confluence with the Wady el Maleh.

The main stream or Wady el Maleh continues for a very

short distance eastward to receive the Wady el Tubkah which skirts the eastern border. Then it runs northward to the confluence with the Wady ed Duba or esh Shukh (alt. 723 feet below the sea).

The Wady ed Duba skirts the eastern moiety of the northern edge of the basin, and drains the area between it and the range of Ras er Raby, which is prolonged by a spur extending eastward between the affluents of ed Duba and the Wady el Maleh.

Below the confluence of the Duba, the Wady el Maleh receives small affluents from the north-eastern part of the basin, near 'Ain el Helweh; and then runs on eastward to the Jordan; which it enters where the plains on the south of Beisan, are terminated by the mountains closing in upon the river.

MINOR BASINS BETWEEN WADY EL MALEH, THE JORDAN, AND WADY FÂR'AH.

The eastern edge of the Maleh basin is only from two to three miles distant from the Jordan. The steepness of the slope towards the valley is best expressed by the actual observations. The river is 1,080 feet *below* the level of the Mediterranean, at the Makhâdet (ford) es S'aidiyeh, and 1,120 feet at Makt. Umm Sidreh. The summits on the eastern edge of the Maleh basin, and on its continuation along the basin of Wady el Bukei'a are as follows:—Ras Nukb el Bakr (alt. 95 feet), Dhahret el Meidân (alt. 653 feet), Kh. Mofia (alt. 590 feet), Ras Umm Zokah (alt. 840 feet), Ras el Jibsin (alt. 110 feet). These elevations above the sea, added to the figures representing the depression of the river below the sea, are equivalent to heights of 1,100 to 2,000 feet above the stream, and they frequently terminate in rocky precipices.

This bold and abrupt slope is broken up by numerous ravines and wadys, rising near the summit, and exhibiting much variety in their descent. The Wady Umm el Kharrubeh runs for three miles parallel and near to the summit before zigzaging downwards to the Jordan. The Wady Shaib has a very oblique course, and it is followed by an ancient road which crosses between the Wady Fâr'ah and the Jordan, at Makt. ez Zakkumeh. On the east of the ford, in a prominent situation, rises the Saracenic castle of Rubûd. About a mile from the river the road to Wady Fâr'ah is crossed by another road which traverses the valley of the Jordan, between Beisân and er Riha, the site of Jericho. South of the Fâr'ah Road, ten other distinct wadys occur along the Maleh slope.

The south-western slope beyond the Maleh basin around the south of Tubas, is drained by the Wady er Resif, an affluent of Wady Fâr'ah. Two roads from Nablûs run in the same direction, side by side along the ridge and furrow of the Resif wady. On a spur from Ras Jadir, descending between two southern branches of Wady er Resif, is Ainûn, which Lieut. Conder considers to represent the Ænon of Scripture (John iii, 23); but although he claims Dr. Robinson's support, the site is rejected by Dr. Robinson, on account of its deficiency of water.* There is a still graver objection to this identification, which will be considered in another work.

Further south, and still at the foot of Ras Jadir, the basin of Wady el Maleh is succeeded by Wady el Bukei'a. Although this wady is of considerably greater extent than the wadys which descend to the Jordan from the eastern edge of Wady el Maleh, it is still only a secondary valley, being cut off by the head of Wady Fâ'rah from the Mediterranean waterparting. Its lower extremity is remarkable, for in approaching the valley of the Jordan, the wady enters a rocky chasm, through which it proceeds southward for more than half a mile, when it doubles back on a serpentine course northward, then north-east and east, through Wady Abu Sidreh, to an offset of the Jordan, at Tell Abu Sidreh.

* Rob., iii, 305, 333. Conder, "Tent Work," ii, 57: "Handbook," 320. Smith's "Bib. Dict.," art. Salem.

The Ghor or upper ground at the foot of the hills, must be distinguished from the Zor or bottom of the valley, about 150 feet lower, in which the channel of the river, cut still deeper, meanders. On the north of Wady el Maleh the Ghor is widened out to the foot of the terrace of Beisân which is about 400 feet higher, and the hills, at first only set back from the edge of the terrace, gradually recede further and further westward, up the valley of Jezreel, and along the Plain of Esdraelon, to Mount Carmel and the Sea.

South of the Wady el Maleh, the hills encroach upon the Ghor, and reduce it to a narrow terrace, which comes to a minimum on the east of Ras Umm Zokah (alt. 840 feet). The Zor also frequently cuts gaps in the Ghor, where the wadys descend into it. This narrowed part of the Jordan Valley extends southward to the Wady Abu Sidreh, when the hills begin to recede westward, and the Ghor again expands, widening gradually (except where the Wady Fâr'ah opens into it), till it acquires its fullest breadth in the plains of Jericho, on the south of Kurn Surtubeh.

Southward between the Wady 'Abu Sidreh and the great Wady Fâr'ah, only one secondary wady can be singled out for notice. It comes from the hills between the Bukei'a and the Fâr'ah, and rises in Ras Umm el Kharrubeh (alt. 690 feet), entering the Jordan about three miles below Tell es Sidreh, with the name of Sh'ab el Ghoraniyeh. About two miles and a half on the south-west of this confluence, the hills on the left bank of Wady Fâr'ah terminate in el Makhruk. The Wady Fâr'ah which has entered the Ghor from the north-west, now takes the name of Wady el Jozeleh, and bends round to the south, meandering in that direction for six miles through the Ghor, to its junction with the Jordan; its distance from the Jordan being only about three-quarters of a mile, nearly all the way. The watersheds here between the Fâr'ah and the Jordan, being thus contracted, leave no room for any other secondary features than mere corrosions in the face of the descent from the Ghor to the Zor. The latter is here remarkable for the remains of the Jisr (Bridge) ed Damieh, with other ruins of the same name, and also for the junction of the Wady Zerka from the east, which is identified with the River Jabbok of the Bible.

The Basin of Wady Fâr'ah.

The western edge of the basin has its northern extremity at a point midway between Tubâs and el Judeideh, the latter being in the inland basin of Merj el Ghuruk, which runs with the Fâr'ah for about three miles. For about a mile and a-half north east of Yasid (alt. 2,240 feet) the Fâr'ah basin runs with that of Nahr el Mefjir, which has a source of its southern branch near Yasid. From Yasid to the south of Mount Gerizim or Jebel et Tor, the western edge of the Fâr'ah runs with the basin of Nahr Iskanderuneh, and passes over Mount Ebal or Jebel Eslamiyeh, and on the east of the town of Nablûs. From Mount Gerizim, the edge of the Fâr'ah basin takes a south-easterly course across the plains, between the plains of Sahel Mukhnah and Sahel Rujib, and reaches the ridge of el Jeddua. Along this part the Fâr'ah basin runs with the Kanah section of the el 'Auja basin. From el Jeddua it makes a bend to the north-east, over et Tuwanik (alt. 2,847 feet) to Sheikh Kamil (alt. 1,920 feet), here the south-easterly course of the edge of the basin is resumed; and continues through Daluk, Umm Halal (alt. 1,360 feet), Ras Kaneiterah, and the noted Kurn Surtubeh (alt. 1,244 feet) to el Mermaleh in the plain which is here depressed 889 feet below the sea, and about three miles south-eastward it reaches the Jordan. From el Jeddua to the Jordan, the Fâr'ah is coterminous with Wady el Humr, which succeeds it on the south. The length of the basin is about thirty miles from north-west to south-east; and its greatest breadth is about twelve miles in its upper part, and about six miles lower down.

The Water Courses and other features of the Fâr'ah Basin.

The western extremity or head of the Fâr'ah Basin is divided into two distinct parts, northern and southern; the latter being a thousand feet higher than the former, in the lowest grounds.

(1.) The southern or more elevated part, includes the continuous Sahels or Plains of Rûjîb, Askar, and Salim. The plain of Rûjîb is a continuation northward of the noted plain of Mukhnah, and it is commonly considered to be a part of the latter. Both are traversed by the high-road between Nablûs and Jerusalem. The only separation between them is the waterparting of the basins of Wady Fâr'ah and Wady el 'Auja, on the south-west of the village of Rûjîb. It is indicated by the commencement of a more rapid descent on the side of Rûjîb.

On the north, the Plain of Rûjîb is succeeded by the Plain of Askar, the biblical Sychar, John iv, 5. The division takes place where the hills recede westwards towards Nablûs, and eastward towards Sâlim, the Shalem of Jacob, Genesis xxxiii, 18; but not the Salim of St. John's Gospel, ch. iii, ver. 23. "Now Jacob's well was there" (St. John iv, 6), and is still, on the southern edge of the plain, just half a mile south of Askar. The plain of Askar is bounded on the north by a range of mountains, an extension eastwards from Mount Ebal, dividing the southern part of the Fâr'ah basin from the northern part, and culminating in Jebel el Kebir (alt. 2,610 feet). The connection of the watercourses on either side of the range, is effected through its intersection by a deep and narrow gorge or chasm named Wady Beidan. The chasm at the entrance from the Plain of Askar, is about 1,500 feet above the sea; but it is only about 600 feet at its exit on the north side of the range, at the foot of Neby Belan (alt. 2,500 feet). The altitude of Wady Beidan is considerably lower, at a distance of a mile and half to the east, where the stream from the mouth of the chasm, joins the waters of Wady Fâr'ah.

Towards the east, the Plain of Askar is followed continuously by the Plain of Sâlim; the division between them being defined by the Wady esh Shejar. The eastern extremity of the Plain of Sâlim is the waterparting between this portion of the basin of Wady Fàr'ah and the basin of Wady Humr, which is the next tributary to the Jordan on the south, and the recipient of the better known Wady Fusail. The altitude of the Plain of Sâlim is 1,500 feet at its western end, and only 1,800 feet at the edge of the descent into Wady Kerad, which is one of the heads of Wady el Humr. But the mountains on the north and south of this end of the plain, rise to 2,510 feet and 2,547 feet respectively. The plain of Askar thus forms the junction of two broad valleys or plains at right angles to each other, and of an equal length of six miles, with an average breadth throughout that seldom exceeds a mile except at the southern and eastern extremities. The Askar plain is the collecting ground of the drainage of this part of the basin before it is carried into the chasm of Wady Beidan to join the lower region in the main valley on the north.

(2.) The northern part of the head of the Fâr'ah Basin is surrounded by a semicircle of hills with a diameter of seven or eight miles. The villages of Asiret el Hatab, Yâsid, Tûbâs, and Tammûn, indicate the course of the circular margin, from which the wadys converge towards two centres. One of them, taking four-fifths of this track, is about half a mile below 'Ain and Tell el Fâr'ah; and the other is at the lower end of Wady Beidan. The inequality of these areas is compensated by the junction with the smaller centre of the outfall of the southern division.

From these central junctions, two streams run, one southward, and the other eastward; and they meet after a course of a mile and a half each. Here the Wady Fâr'ah begins its south-eastern course to the Ghor, with the summits on each side about four or five miles apart. After a descent of three miles and a half, the valley is found by the observations of the survey to be on a level with the sea. At Yasid on the western edge of the basin, and eight miles and a half distant, the altitude is 2,240 feet. At the junction of Wady Fâr'ah with the Jordan, the depth is 1,160 feet below the sea level, the direct distance from the sea level point being about fifteen miles, which by the winding of the river is increased to not less than eighteen miles. The fall from Yasid to the sea-level-point in Wady Fâr'ah, is therefore about 264 feet per mile, or exactly one in twenty; while the fall from the same point to the Jordan is only about 77 feet per mile or about one in sixty-eight.

The Wady Fâr'ah below the junction of its head waters, presents three natural divisions of about equal length, according to the variation of its landscape. In the uppermost part, the river flows through a beautiful basin of meadow land with the stream flowing in the midst bordered by oleanders.

In the central part the river descends chiefly amidst precipitous rocks, which here separate its bed from the fertile slopes above. Towards the lower end of this part, on the right bank, occurs a beautiful tract which descends to the river bank, where it is covered with oleanders. It is called el Fersh, and by Dr. Robinson "Fersh el Musa," "Bib. Res." iii, 304. On the left bank are the ruins of an ancient town now named Buseiliyeh, visited by Van de Velde, Guerin, and the Surveyors of the Palestine Exploration Fund. The central part is terminated by the projection of a spur of the hills on the north, met by precipitous rocks on the south, which close in upon the river, and reduce its passageway to a narrow gorge, which is remarkable for its caverns and the colour of its rocks.

The lowest part begins below the gorge, and spreading out over the marshes of the Kurawa, extends to the Ghor. Van de Velde describes the Kurawa as a "well watered and richly wooded oasis, with luxuriant fields and gardens, and oleander-bordered brooks." It is the principal encampment of the Mas'ûdy Arabs. Ruins only now remain of mills and houses, where once stood the city of Archelaus.

But few affluents from either side occur in the upper and central parts of the valley, of sufficient importance to call for special notice. At the higher end of the upper part, and on the left bank, the Shab esh Shinar descends very obliquely from the northern slope of Jebel Tammun and the village of the same name. At the lower end of the same part, a wady descends from a pass (Nukb el Arais) which facilitates communication with Wady el Bukeia, and divides Jebel Tammun from the heights of Jurein, Homsah, and Kharrubeh. These heights are broken and precipitous, more so than those on the opposite bank, which although terminating in cliffs along the course of the river, descend to them by broader slopes. The cliffs on both sides of the stream distinguish the central part. The Kurawa receives all its notable affluents from the right bank, including Wady ez Zeit, Wady Jabr, Wady el Khurzeleiyeh, and the Talat el Kurein from Kurn Surtubeh. Below the Kurawa, the Wady Fâr'ah takes the name of Wady el Jozeleh, which has been already noticed.

Dr. Robinson remarks that the Wady Fâr'ah (Fâri'a) is "justly regarded as one of the most fertile and valuable regions of Palestine," "Bib. Res." iii, 304. Being subject to the nomadic Mas'ûdy Arabs, it is without villages, except on the western margin of the basin. But it abounds with pastures and cornfields, and supports large herds of cattle and quantities of goats. "Nowhere in Palestine had I seen such noble brooks of water," exclaims Dr. Robinson ; and Mons. Guerin expatiates on the delicious shade of gigantic fig trees, the magnificent shrubs and beauteous oleanders, which line the banks of the streams, Guerin, " Samarie " i, 258. "A most delightful place," "knee-deep in beautiful flowers,"—"this charming valley,"—are among the praises heaped upon Wady Fâr'ah by Lieutenant Conder, " Tent Work " ii, 57.

Important roads intersect Wady Fâr'ah in various directions. The great north road from Jerusalem to Nazareth, Beisân, the Sea of Galilee, and the regions beyond, cross the western head of the basin in its widest part. Several main highways to Gilead and the east of the Jordan—coming from Tubâs, Sannur, Yasid, Tulluza, and Nablûs,—meet in the Wady Fâr'ah, and then pursue a common route to the Jordan at the ford of Damieh ; various cross tracts will also be found on the map.

As " Abram passed through the land unto the place of

Sichem," Gen. xii, 6, he probably followed the road across the head of the Fâr'ah Basin. Jacob returning from Laban by way of Mount Gilead with his wives, children, and servants, his herds of cattle, flocks of sheep and goats, camels, and asses, ascended the Fâr'ah from the Jordan, and "came to Shalem, a city of Shechem," Gen. xxxiii, 18. Benhadad the Syrian, fleeing from his siege of Samaria, panic-stricken by the Almighty, hurried down the Fâr'ah Valley, " and lo! all the way was full of garments and vessels which the Syrians cast away in their haste," 2 Kings vii.

THE BASIN OF WADY EL HUMR.

This is the Wady el Ahmar of Van de Velde, " Sinai and Palestine," ii, 315; "Memoirs," 123. Robinson calls it Wady Ahmar, "Bib. Res." iii, 294. It is better known in connection with Wady Fusail, a minor branch of the basin, containing the site of Phasaëlis. Its general outline may be compared to a right angled triangle, with the southern boundary for its base, dividing it from the basin of Wady el 'Aûjah which falls into the Jordan next to this on the south. The southern boundary passes from the Jordan through el Arâka, Kh. Jibeit (alt. 2,146 feet), and el Mugheir (alt. 2,246 feet), to the eastern edge of Merj Sia, a small natural basin with no outlet. The length is about eleven miles.

The perpendicular of the triangle forms the western boundary, running north and south with some small sinuosities; on this side also the basin is coterminous with the 'Auja basin, but here it is quite another 'Auja from the 'Aûjah on the south, and is indeed the great basin of Nahr el 'Auja, which falls into the Mediterranean Sea on the north of Jaffa. The slight difference in spelling may be unintentional. The western boundary is traced from Merj Sia, along a ridge between Istuna and Kulason, and east of the sources of Wady Seilun, then to about midway between Jalûd and Domeh, and onwards to a point nearer to Kusrah than to Mejdel Beni Fâdl; further north it crosses Akrabeh, and passes north-west of Yanun, to the mountain of el Jeddua and et Tuwanik (alt. 2,847 feet), then near Tana to the northern extremity of the basin near Sheikh Kamil (alt. 1,923 feet). The length is about twelve miles.

The hypotenuse of this triangular basin faces the northeast and is coterminous with the Wady Fâr'ah, in connection with which it has been already traced. Its length is about thirteen miles.

The Watercourses of Wady el Humr Basin.

Two main streams receive the channels of this basin, and unite about a mile from the Jordan. These are the Wady el Humr which drains the northern part of the basin, and the Wady Fusail which is the outfall of the southern part. There is an intermediate channel which rises about two miles from the edge of the plain, and passes straight across it to the confluence, a further distance of five miles.

The sources of Wady el Humr extend along the western waterparting for about eight miles, between Sheikh Kamil and Mejdel Beni Fâdl, and they form two divisions. The first rises at Tana and runs with the Roman road as Wady el Kerad into the Sahel or Plain of Ifjim. It receives a branch from Sheikh Kamil at a point midway between Tana and the plain; and another branch comes from the same range of hills through Lahf Salim, and joins the Kerad at the upper end of the plain of Ifjim. At the lower end of the plain the Wady Zamur joins the Kerad. The Zamur has the name of Wady ed Dowa above its entrance into the plain, and is the recipient of a series of tributaries which severally rise at Kh. Yanun, el Jeddua, Yanun, Akrabeh, and north of Mejdel Beni Fâdl. These drain an upland tract, enclosed between the western waterparting and spurs which proceed from it and are drawn together at the gorge of Wady ed Dowa, the streams having united at the entrance of the gorge. After the junction of the Zamur the wady takes the name of Wady el Ifjim and receives short branches from the north-east parting at Bir Abu Deraj, Umm Hallal (alt. 1,360 feet), Ras el Hufireh, and

Ras Kuneiterah. It receives longer branches on the opposite or right bank, descending from the eastern face of the spur already mentioned, which extends from the western parting at Mejdel Beni Fâdl northwards to Wady ed Dowa. These are named Khallet 'Aseim, W. el Menakhir, called also es Subhah, Wady Abu Hummam, and Wady Saddeh. .

The Wady Ifjim proceeds towards the Ghor from the north; but before reaching it, a deflection takes place which causes the wady to pass into the plain through a precipitous chasm from the west, which meets at its entrance from the plain, a similar chasm coming from the north, as if it had been the passage of the Ifjim before some convulsion diverted the stream to the western chasm.

At its entrance into the Ghor the wady is called Zakaska, and runs at the foot of lofty rocks on its right bank, while the slopes of Kurn Surtubeh have their base about a mile off on the left, and finally reach their southern extremity where the wady proceeds westward across the Ghor, as Wady el Humr. Its junction with Wady Fusail takes place in the low ground of the Zor, the descent to which is here less abrupt than usual, owing to the channels having worn down the surface of the Ghor, giving it a rough and broken aspect, for a considerable distance from the river.

The Wady Fusail has its sources on the western water-parting between Mejdel Beni Fâdl and Merj Sia. The Wady Bursheh running eastward on the south of Mejdel Beni Fâdl, receives two small feeders from the west, and two from the east of that village, and then goes to the south-east, receiving the Wady Arak esh Shaheba from the village of Domeh (alt. 2,000 feet), and continuing in the same direction for three-quarters of a-mile lower down, as the Wady Arak Hajaj. Here it receives the Wady er Rishash from the south-west, the numerous sources of this branch being spread out along the south-western margin of the basin, between Domeh and el Mugheir (alt. 2,245 feet).

After the junction of the Rishash, the Arak Hajaj proceeds eastwards for a mile and a-half. Then entering the

Ghor it takes a south-easterly course across the plain of Wady Fusail, and reaches the confluence with Wady el Humr, through Melahet Umm 'Asein. As the Fusail deflects from the foot of the hills, it receives from them Wady el Makthayeh, Wady Abu Zerka, and Wady el War, also Birket Fusail and another fountain in the plain, amid aqueducts and ruins that denote the site of the ancient Herodian city of Phasaëlis.

The Basin of Wady el 'Aûjah.

The shape of the margin of this basin may be compared almost to an ellipse or to a rhomboid, with the four sides bulging outwards, the two longer being on the north and south. The parallel inclination of the shorter sides in passing from the northern border southward, is slightly to the east. The eastern waterparting follows the course of the Jordan at a distance from the river of about one mile and a-quarter at the northern end, tapering to half-a-mile at the southern end, where the basin has its outlet into the Jordan. This narrow ridge is about seven miles in length, and its summit being on a level with the Ghor, denotes its identity with that feature, from which it is only separated by the gradually declining course of the Wady el Mellahah, to join the 'Aûjah, near the outlet of the basin into the Jordan, where the depression below the sea is 1,200 feet. The Ghor seems to be here about 400 feet higher.

The northern waterparting concurs with that of Wady el Humr as far as Merj Sia, and this part is described in the account of that basin. But it is prolonged further westward for about three miles so as to include Kh. Abu Felah. Here the western boundary begins, running south to Tell Asûr (alt. 3,318 feet). From Merj Sia to Tell Asûr, this basin impinges on the Mediterranean waterparting of Nahr el 'Auja. Southward of Tell Asûr, the Mediterranean system trends south-west, and the boundary of this basin trends south-east, following the Roman road as far as Kubbet Rummâmaneh

WADY EL 'AÛJAH. 85

(alt. 2,024 feet), allowing the head of the Nûei'ameh, the next Jordan basin, to intervene. From Kubbet Rummâmaneh it runs eastward to Umm Sirah, where it passes to the southeast to make a precipitous descent into the Ghor, on the north of 'Ain ed Dûk and 'Ain en Nûei'ameh, from whence it bends round to the north-east by 'Osh el Ghurab, Maidan el Abd, Khurbet es Sumrah, and the Jordan on the south of the outfall at el 'Aujah.

The Watercourses of el 'Aûjah Basin.

Three divisions of this basin may be distinguished, namely, Wady el Mellahah in the northern part; Wady el 'Aûjah in the centre; and Wady Abu Obeideh in the south; with their respective affluents.

Wady el Mellahah originates in a long swamp at the northeastern extremity of the basin, and running along the eastern margin, joins el 'Aûjah near the outfall. At the upper end of the swamp, it receives Wady Unkur edh Dhib, which rises on the south of Kh. Jibeit (alt. 2,146 feet), and skirts the northern margin of the basin. Two wadys with parallel courses to edh Dhib enter the swamp lower down; and two more, including Wady Bakr, flowing in a similar direction, enter Wady Mellahah after it leaves the swamp. The Wady Mekûr edh Dhib, on the south of Wady Bakr, is dispersed by irrigation channels in the Ghor, otherwise it would contribute to Mellahah.

From Wady Zakaska, where the Wady el Humr enters the Ghor, to Wady Bakr, the descent of the mountain side, at first precipitous, continues steep, and in the same line north and south. But on the right or south bank of Wady Bakr, the base of the mountains begins to be extended in the form of low hills for a mile and a-half eastward, and continues so southward to the Wady el 'Aûjah.

South of Wady el 'Aûjah, these hills are separated from the mountains by a plain (the Emek or Plain of Keziz Joshua xviii, 21), until they reach their southern limit, and

G

approach the rocky cliffs and precipices of the mountain base about Jebel Kuruntul, where the Wady Nûei'ameh intervenes.

They are intersected by the Wady el 'Aûjah, and also by the Wady Abu Obeideh, called also Abideh, probably by an oversight. Between the Wadys el 'Aûjah and Obeideh, the hills throw out a long, low, and narrow tongue across the Ghor. At el M'adhbeh, they attain to an altitude of 283 feet above the sea, the mountain of en Nejmeh on the west having the alt. of 2,391 feet. The Ghor at Kh. es Sumrah, at the eastern foot of el M'adhbeh, is 840 feet *below* the sea, while the enclosed "Plain of Keziz" between the mountains and the hills, is about 200 feet below the sea.

The Wady el 'Aûjah has its principal sources in the north-western extremity of the basin, and receives several tributaries from its margin between el Mugheir and Tell Asûr. At 'Ain Samieh it acquires the name of that source, and enters the rocky defile by which it proceeds to the enclosed plain, that it has been proposed to identify with the Benjamite settlement of Emek Keziz. About a mile and a quarter before leaving the mountains, it takes the waters of 'Ain el 'Aûjah, and becomes a permanent stream with that name. After crossing the enclosed plain, it enters the hills at the northern foot of el M'adhbeh, and receives in the gorge, the Wady Abu el Haiyât on the left bank, and the Wadys Sebata and el Abeid on the right bank. The two Wadys el Haiyât and Sebata only rise on the outer slope of the mountains; but the Wady el Abeid takes the waters of Wady en Nejmeh, which descends from Mount en Nejmeh (alt. 2,391 feet); also those from the deep and rocky chasms of Wady Dar el Jerir and Wady Lûeit.

The Wady Dar el Jerir comes from the highland villages of **Kefr Malik** and Dar Jerir, on the eastern slopes of Tell Asûr (alt. 3,318 feet). It is the Wady Habis and Wady 'el Musireh of former maps. The Wady Lûeit in its upper course is called Wady et Taiyibeh, and descending from near the village of that name, the Ophrah of Scripture (alt. 2,850 feet), skirts the southern margin of the basin until it approaches the opening of its chasm into the plain. It was confused formerly with Wady Habis and Wady el Musireh. The Wady Abu Obeideh has its source on the west of Umm Sirah, and takes the rest of the drainage of the southern border of the basin. In crossing the enclosed plain, here called Salet el Meidan, the Wady Abu Obeideh receives the Wady Umm Sirah, which rises among the rocks at the foot of the mountains near the pass of Nukb el Asfar. This is the Wady el Musîreh of former maps, which confused it with Wady Dar el Jerir and Wady Lûeit, and carried it into the Wady el Aûjah, instead of into Wady el Obeideh. It also receives the Shukh ed Dub'a, and then begins to cross the low hills on its way to the Ghor, and to its junction with Wady el 'Aûjah in the depths of the Zor.

THE MINOR BASIN OF WADY MESÂ'ADET 'AÎSA.

A small group of secondary basins succeeds the el 'Aujah, and intervenes between its outfall and that of Wady Nûei'a- meh. The only notable one amongst them is Wady Mesâ'adet 'Aîsa or " the Ascension of Jesus," with several small branches, which drain the eastern side of the southern part of the detached hills, from the Maidân el 'Abd to the 'Osh el Ghurab or " Raven's Nest " a traditional Mountain of the Temptation of Our Lord, from whence the name of the Wady is derived. (Conder's " Tent Work," ii, 5, 10, 13.) This tradition is said to be only attached to the 'Osh el Ghurab at the present day by the Bedawîn; but as Lieut. Conder also attributes to it a " mediæval monkish " origin, for which he cites authorities, it may be observed that the opposite summit of the Kuruntul* Mountain is reputed by the Roman Church to be the " exceeding high mountain " of the Temptation. Le Frère Lievin, " Guide des Sanctuaires," 377.

* Called also " Quarantania," or Mount of the Forty days' Fast.

88   THE JORDAN WATERSHED.

THE BASIN OF WADY NÛEI'AMEH.

This narrow basin seldom exceeds three miles in width, and it is confined to barely half a mile in its lower course. Its sources, rising on the Mediterranean waterparting, at a distance of about twenty miles from its junction with the Jordan, lie between Tell 'Asûr (alt. 3,318 feet) and the well-known village of Beitin or Bethel (alt. 2,890 feet). The highroad from Bethel to the North, runs along the waterparting for about four miles; and another road follows near it for the rest of the distance to Tell 'Asûr, or about three miles.

The curvature of the northern boundary has been described in the account of the El 'Aûjeh basin as far as 'Osh el Ghurab; where the interposition of the preceding secondary basin causes the present boundary to bend to the south-east, passing Kh. el Mefjir, and then east along the left bank of Wady Nûei'ameh.

The southern boundary, starting from Bethel, follows the road to Deir Diwan (alt. 2,570 feet), and southward to Mukhmas (Michmash) (alt. 1,990 feet), and Ras et Tawil (alt. 1,964 feet), whence it proceeds eastward along a mountain track to Umm et Talah, and Jebel Kuruntul (alt. 320 feet), on the north of which the track descends to the Ghor by a gap in the line of cliffs, and passing on the north of 'Ain es Sultan, follows the right bank of Wady Nûei'ameh to the Jordan which is here 1,230 feet below the sea level. This boundary has a general curvature parallel with that on the north of the basin, and deflecting in a similar manner from the Mediterranean waterparting, so as to interpose between it and the present basin, the north-western part of Wady el Kelt, which falls next into the Jordan on the south. Thus while the north-western part of this basin impinges on the south-eastern part of the el 'Aûjeh basin belonging to the Mediterranean slope, the south-western part is divided by a portion of the Wady el Kelt basin from the famous plain of el Gib (Gibeon) or Neby Samwil, which occupies the northeast of the Nahr Rubin division of the Mediterranean watershed. The effect of these curvatures is to facilitate lateral communications parallel with the main range or axis of the mountain system of the country. Facilities of the same kind are also provided occasionally by the course of the head streams, when they run parallel with the main range, and sometimes come from opposite ends of the same valley, before they unite to make a rectangular or an oblique descent to the lower grounds.

The Watercourses of the Nuei'ameh Basin.

The Wady el 'Ain descending southward from Tell 'Asûr (alt. 3,318 feet) receives on its right bank from the western edge of the basin,—the Khallet es Sultan, Wady el Kanabis, and Wady Muheisin. A branch having the villages of Dar Jerir and et Taiyibeh on the east, joins the left bank, about a mile south-west of the latter village. The Wady Muheisin runs eastward along the northern foot of the ridge between Beitin and Deir Diwân. After its junction with Wady el 'Ain, the course of the Wady Muheisin is continued southeastward, and this name is changed to Wady Asis. The wady here enters a deep and rocky chasm in which it continues for five or six miles. A tributary from the west of Rummôn ("the rock Rimmon," Judges xx, xxi), running south for about 1½ miles, joins the main Wady at a mile from Wady el 'Ain. Another tributary rising on the south of Taiyibeh, where it is called Wady Abu el Haiyat, passes southward on the east of Rummôn as Wady el Asa, and after a course of three miles meets the wady from the north-west. Not far below a wady falls in on the right bank, which rises on the hills on the east of Deir Diwan and Kh. Haiyan (Ai). After this junction the wady trends slightly east of south for a mile, and then bends to the east, and receives the Wady es Sineisileh from the border of the basin near Kubbet Rummâmaneh (alt. 2,024 feet). It continues eastward in the chasm as W. Rummâmaneh, receiving the Wady el Harik* on the

* The Wady Harik stands in name only for the Wady Harith of former authorities (Stanley's "Sinai and Palestine," 201).

right bank, and retaining the former name till it receives the Wady el Makûk on the same side. The long rocky chasm ceases about half a mile before the junction with the Makûk. The Wady el Makûk rises midway between Deir Diwan and Mûkhmas, near the road that connects those villages, and passes Ras et Tawil (alt. 1,964 feet) in a rocky chasm called W. Sikya. It is the drain of the south-western part of the basin. After the confluence the Makûk is bulged slightly to the northward by el Ṭubakah, a spur from the southern waterparting, which provides a tributary from the valley on its southern side. The Wady now dives north-eastward into a rocky chasm, where it receives the Wady Abu Jurnan, which rises near Khubbet Rummâmaneh (alt. 2,024 feet) and skirts the northern edge of the basin. The chasm conducts the wady to the southern extremity of the enclosed plain, where 'Ain en Nûei'ameh supplies its final name as well as a perennial stream, the latter being also augmented by 'Ain ed Dûk. The Nûei'ameh now flows to the south-east, and divides the southern extremity of the hilly tract which terminates in 'Osh el Ghurab or the Raven's Nest, from the long line of lofty cliffs which here forms the eastern base of the Mountains of Judæa. It crosses the Ghor on an easterly course, and descends to the Jordan at the ford of el Ghoraniyeh (1,250 feet below the sea).

THE BASIN OF WADY EL KELT.

The western edge of this basin is about nine miles in length, beginning on the north at Bethel. As far as Bireh, it joins the el 'Auja basin, which falls into the Mediterranean on the north of Jaffa. From Bireh to Shafat the western edge meets the north-eastern part of the Nahr Rubin basin, which includes the Plain of el Jib or Gibeon on the north of Jerusalem, and which enters the Mediterranean on the south of Jaffa.

The northern edge has been already described (p. 84). The southern departs from the west near Shafat; bends round to Anata, and eastward to the rock of Arak Ibrahîm; then it makes another sweep north-east and south-east to Khan Hathrurah on the Jerusalem-Jericho road; then advances by Talât ed Dumm to Kh. el Mestrab, and continues along the ridge between Wady el Kelt and Wady Talat ed Dumm, descending to the Ghor between Khaur Abu Dhahy and Khaur et Tumrâr, passing south of Ain Hajlah and reaching the Jordan at the Pilgrims' Bathing place or Makhadet Hajlah.

At the western margin of the basin, its width is about nine miles. Where the head waters descending from it unite, —at the confluence of Wady Suweinit from the north-west, with Wady Fârah from the south-west,—the width of the basin contracts to three or four miles. At the foot of the cliffs that form the base of the mountains in the plain of Jericho, the width is about two miles and a half, and in the eastern part of the Ghor it is about a mile. The length reckoned from Bethel to the Jordan is about twenty-three miles, or nineteen miles direct from Tell el Fûl. The Kelt is the southernmost affluent of the Jordan from the west.

The Watercourses of Wady el Kelt.

The watercourses rising on the margin of the broad head of the basin are divided into two parts. The northern part contributes to Wady Suweinit, and the southern to Wady Fârah. These unite in the Wady Kelt, about midway between the western waterparting and the foot of the mountains.

The most northerly sources of the Suweinit are two brooks on the south of Bethel, which soon unite at the foot of Kh. Ibn Barak, where a third also falls in from the north of Bireh. From the junction, the wady runs south-eastward, till it joins another wady on the east of Burkah, coming from the north. This wady from the north, is the recipient of three parallel branches also running south-eastward, which rise ôn the south side of the road along the waterparting between Bethel (alt. 2,880 feet) and Deir Duwan or Diwan (alt. 2,370 feet). These branches are divided by spurs from the waterparting, the spur nearest to Deir Diwan having at its extremity about half-a-mile south of Deir Diwan the ruins of Kh. Haiyân, which the successive observations of Robinson, Guerin, and Conder have identified with the city of Ai taken by Joshua. See note on Ai, p. 95.

The wady from Kh. Haiyân appears to derive from that ancient site, the name of Wady el Medineh or the Valley of the City. From the confluence east of Burkah, it turns south-westward to a small plain on the south of that village, where it receives two affluents from the east and south of Bireh (alt. 2,820 feet). It runs on to the south-east for two miles, till it is joined by Wady en Netif, which has its sources on the waterparting from Kefr Akâb to er Ram (alt. 2,600 feet), and passes on the north of Jeba ("Geba," 1 Sam. xiv). The main wady proceeds from the junction for three-quarters of a mile to the east and north-east, up to the entrance of the long, narrow, and rocky gorge, of Wady es Suweinit. At the beginning of the gorge, a wady falls in from the north-west, after receiving a short branch from Mukhmas (biblical Michmash), which is on the northern waterparting within a mile north of the gorge. Lieut. Conder in his "Tent Work," ii, 112-115, seems to place the Philistine camp which Jonathan seized, on a tongue of land, coming to a sharp point between Wady Suweinit and another gorge that joins it on the east of Kh. el Haiyeh and, from one to two miles south-south-east of Mukhmas. According to him the southern face of this point is the rock Bozez or Shining; and the opposite side of the gorge facing the north is the rock Seneh, meaning the thorn or Acacia, the present name of the valley being Suweinit, or the Little Acacia.

From the head of the Suweinit Gorge, to its junction with the gorge of the Fârah, which drains the southern division, the course of the Suweinit is south-east, and its length is about four miles. About three-quarters of a mile from the head of the gorge a small branch falls in from the Jeba plain on the west. The next branch, joining on the opposite bank about a mile lower down, forms the tongue of land identified with the Philistine camp before mentioned. Three other branches join the main stream on the same side within a mile of the junction with the Fârah. They descend from Ras et Tawil (alt. 1,864 feet) on the northern edge of the basin.

The Wady Fârah takes the drainage of the western water-parting between Khurbet Erha (alt. 2,450 feet) and Shafat (alt. 3,524 feet). The head wadys fall into the Wady Fârah by two branches, the Wady Redeideh coming down on the north of Hizmeh, and the Wady es Senam on the south of the same place, the junction being in a precipitous gorge one mile east of Hizmeh (alt. 2,020 feet). The easternmost branch of. the Redeideh descends from Jeba, which has the Wady en Netif on the north.

Wady es Senam, receives the Wady Zimry from the east of Tell el Ful (alt. 3,754 feet). Another branch comes from the south side of the same prominent hill, taking a south-easterly course along the southern edge of the basin, till it approaches Anata, the biblical Anathoth (alt. 2,225 feet), and then bending abruptly to the north and north-east, it falls into Wady Zimry, on the south of Hizmeh.

After the junction of the Redeideh and Senam, the gorge of the Fârah proceeds due east with a slightly serpentine course to its junction with Wady es Suweinit, receiving the Wady en Nimr from the Plain of Jeba on the left bank, and the Wady en Nukheileh on the right, the latter having skirted the southern edge of the basin from the neighbourhood of 'Anâta to its junction with the Fârah, within a mile of the Suweinit.

Below the meeting of its two main branches, the gorge of Wady Fârah turns to the north-east, then east and south, and again north-east and south-east, when it emerges for a time from the cliffs at 'Ain el Kelt, and becomes Wady el Kelt. Proceeding eastward, the Wady el Kelt soon receives Wady Abu Duba from sources on the southern edge of the basin, where the Khan Hathrurah and Talat ed Dumm occur on the road between Jericho and Jerusalem. About two miles and a half before it reaches the plain of Jericho the Kelt again enters a deep and rocky defile, which skirts the southern edges of the basin, and continues to the line of cliffs that form the base of the mountains on the west of Jericho. The wady proceeds due east across the Ghor as far as Eriha, the modern Jericho, situated within two miles of the cliffs; after which it turns south-eastward, and soon receives a perennial affluent from Tell es Sultan in the midst of the vestiges of an ancient site of Jericho. The Khaur Abu Dhaby next falls in on the opposite bank, its origin being due east of its outfall, on the mountain side. The last tributary rises in Ain Hajlah, a little more than two miles from the Jordan and unites with the Kelt at the edge of the Zor.

Wady Rijan.

Between the Nûei'ameh basin and Wady el Kelt, there is a wady which descends from Ras et Tawil (alt. 1,864 feet), and passes eastward to the plain, on the south of Jebel Kuruntul. In the plain it appears to be exhausted in irrigation, but naturally it seems to belong to the Kelt. An old road comes down through it from Bethel and Mukhmas to 'Ain es Sultan, but on approaching the plain the road ascends the ridge on the south of the wady, and turns to the south for nearly a mile along the top of the cliffs, before it descends to the plain. This wady has the following names in succession, Shamut, Rijân, el Mefjir, and Abu Retmeh. Besides the road already mentioned, there is another which runs parallel to it on the north, till both meet at Deir Diwan. This track ascends from the plain on the north of Jebel Kuruntul, and follows the ridge between Wady Rijan and the Nûei'ameh basin, keeping about half a mile or a little more from the other route, all the way to Deir Diwan. On the north of Ras el Tawil, a third parallel route begins, which also runs on to Deir Diwan. The central route follows the upper course of Wady Sikya or Makûk, in the Nûei'ameh basin, and the routes on either side traverse the hills which enclose it, the western passing along the waterparting between the Nûei'ameh and Kelt basins, and through the village of Mukhmas. In the direction of these roads eastward from Mukhmas was probably the Hyenas' Ravine, called in the Bible the Valley of Zeboim— "The way of the border that looketh to the Valley of Zeboim toward the Wilderness," 1 Sam. xiii, 18. "The border" in this passage cannot refer to the border between Judah and Benjamin, as it has been suggested, because the locality is in the central part of Benjamin. The original admits of being rendered thus—"the way of the edge (or ridge) overlooking the Hyenas' Ravine in the wilderness." The Ravine of Wady Sikya would answer to this rendering, and if the Upland pastures of Makûk, Rijan, and Kelt, offered no plunder in those days, then the expedition would have had the plain of Jericho for its destination, as it was quite within reach.

Note on Ai, supplementary to page 92.

The name of Khurbet Haiyân is given to the remarkable ruins on this site by Lieut. Conder. Mons. Guerin calls them Kh. el Koudeireh, and describes them fully (Guerin, "Judée," iii, 57–62). Van de Velde's enlarged map of the Environs of Jerusalem has the same name. It appears on the Survey on the west of Haiyan, and quite apart from that. Dr. Robinson, who discovered these ruins on the 4th of May, 1838, does not supply any modern name to them. He first mentions their identification with Ai doubtfully, as "the site with ruins south of Deir Diwân" ("Bib. Res.," i, 443). Returning to the same place ten days afterwards, he examined it carefully and recorded his measurements and notes. After further search he states that he "could come to no other result, than to assign as the probable site of Ai, the place with ruins just south of Deir Diwan." "Bib. Res.," i, 573, 4, 5. Lieut. Conder has repeatedly included this site among his own identifications ("Handbook," 254, 402; "Tent Work," ii, 109; "Biblical Gains," 5, 20,) without any reference to Dr. Robinson's prior claim, so that some reminder seems necessary.

The light thrown by the Survey on this interesting subject, is too explicit to be passed by on this occasion. The two wadys that flank the sides of Haiyân, correspond to those on the north and west of Ai specified in Joshua viii. The wady on the north was interposed between Ai and Joshua's prominent position on that side of the city, which appears to have been just where Deir Diwan now stands (Josh. viii, 11). The other wady on the west was the place of "ambush." It is a defile stretching up for a mile and a half towards Bethel "between Bethel and Ai on the west side of the city," Josh. viii, 12. The road between the two places runs along the summit of the north side of the defile. There is a more secluded parallel valley on the south, which may also be said to be on the west of Haiyân. The King of Ai appears to have made his attack upon the Israelite camp on the north of the city, "before the plain" or literally "on the face of the Arabah;" and the Israelites made their feigned flight by the roads to the Wilderness (midbar) on the south-east of Deir Diwan. This was probably the line by which they had been routed on their first attack on Ai; and the chase unto Shebarim (Josh. vii, 5) may perhaps be commemorated in Kh. Abu Sabbah, although that name is said to be derived from a family who once resided there. Abu Sabbah lies on the line of the route, about a mile from Haiyân. The "going down" or Morad, along which the pursuit was maintained may be traced by the track to the northern foot of Ras et Tawil, and eastward along the ridge Umm et Talah, to the descent on the north of Kuruntul Mountain, down to the plain of Jericho. This line seems preferable for the purpose, to the "ancient road" through Mukhmas (Michmash) and Wady Rijan, which is perhaps of later date.

This subject formerly drew from Dr. Stanley a dissertation on the topography of the mountains in relation to it, which afforded all the light that could be thrown upon them, up to the publication of the New Survey. "Sinai and Palestine," 201–203. His Wady Harith seems to derive its name from the Harik of the Palestine Exploration Survey, corresponding to the Harit of Van de Velde's map. But it is hard to say whether the valley itself was meant for the Muheisin at the head of the Nûei'ameh basin, which is the conclusion suggested in the first line of page 202, for the last line of the same page, points to Wady el Medineh, as descending to W. Kelt.

THE BASIN OF EL KUEISERAH AND THE MINOR BASINS ON THE NORTH.

The next basin adjoining the Mediterranean waterparting falls into the Dead Sea at a direct distance of more than five miles on the south-west of the junction of Wady el Kelt with the Jordan. Near its outlet at the foot of the mountains, it bears the name of el Kueiserah in the new Survey. It is called Wady Dabor in the maps of Canon Tristram and Lieut. Van de Velde, and Wady el Kuneiterah in the map of the Holy Land in Dr. Wm. Smith's Atlas, the last name being derived from the mountain on the south of its exit into the plain, and it is so called by Dr. Wilson. Its contact with the Mediterranean waterparting is very slight, being confined to about one mile on the south of Shafat, where it meets the Wady Beit Hannina in the Basin of Nahr Rubin. Further south it is bounded on the west, by the northernmost part of the basin of Wady en Nar or Brook Kidron, where the City of Jerusalem is situated.

The northern boundary has been already described as a portion of the waterparting of Wady el Kelt, as far east as Khan Hathrurah. Further east this basin is divided from the Kelt by the Wady Talat ed Dumm, and other secondary basins.

From the head of Wady Talat ed Dumm the northern boundary of the Kueiserah Basin divides it from the secondary basins by passing south-eastward to Jebel Ekteif and Neby Musa, where the waterparting descends to the plain and through Belawet edh Dhehaiban to the Dead Sea. The description of the secondary basins will now be taken up.

98   THE DEAD SEA WATERSHED.

Wady Talat ed Dumm and other Minor Basins of the Dead Sea.

This wady is traversed throughout by the road between Jerusalem and the Jordan, the road to Jericho also only passing from this valley into the gorge of the Kelt, about two miles from the plain. Shortly before it descends to the plain, the Wady Talat ed Dumm changes its name to Wady Medhbah Aiyad. In crossing the Ghor it receives the Wady el Hazim from the lower part of the hills and taking the name of Khaur el Tumrar, it passes south-eastward to the Zor, then southward through the Zor and parallel to the Jordan to its outfall into the Dead Sea.

Wady Makarfet Kattum.

About half a mile west of this outfall, the Dead Sea receives another independent wady from the foot of the hills, named Makarfet Kattum.

Wady Joreif Ghuzûl.

A mile still further west, a third wady falls into the sea, which has its origin on the west of Jebel Ekteif (alt. 640 feet) about four miles from the plain, and crosses the plain as Wady Joreif Ghuzûl.

The southern boundary leaves its brief contact with the Mediterranean Waterparting at a point on the road between Jerusalem and Shafat, about a mile on the south of the latter place; and follows the ridge to the summit of the Mount of Olives. It descends from the summit eastward and then southward, leaving el 'Azirîyeh and Abu Dis on the left, and continues southward to Kh. Jubb er Rûm. Soon after it turns to the east straight along an ancient road, until a track falls in from Wady Abu Hindi, near which it is deflected northward for a short distance, and then bends again eastward to reach the summit of el Muntar* (alt. 1,723 feet). From this elevated point it turns north-east along the range which

* Conder's "Tent Work," i, 290, 300, 301.

divides Wady Dekakin from el Bukei'a and Wady Kumrân, until it reaches el Hadeidûn* and approaches Wady Mukelik, when it proceeds eastward across the northern extremity of the plateau of el Bukei'a to Tubk el Kaneiterah (alt. 306 feet, or 1,598 feet above the Dead Sea). Here the boundary of the basin strikes northward to the summit of Jafet el Asla, and thence bends east by south across the plain to the Dead Sea.

The Watercourses of el Kueiserah Basin.

A Wady rising at Shafat on the north-west angle of the basin, runs under various names, along the whole of its northern margin to the Dead Sea. As the Wady Seleim it rises near Shafat, passes on the north of Aîsâwîyah and south of Anâta to the gorges of Deir es Sidd, where it receives the Wady Ruabeh coming from the south of Anâta, and the eastern slope of the northern extension of the Mount of Olives. Below Deir es Sidd, it appears to take the name of Wady es Sidr, and runs eastward till it is diverted to the southward by the waterparting at the head of Talat ed Dumm. A road from Jerusalem passes on the north of the Mount of Olives, and after being joined by another from the summit, skirts the Wady Ruabeh and the Wady es Sidr, till it crosses over to the Talat ed Dumm as already mentioned. The pass is called Thogret ed Debr in the Survey; and to this, the name of Wady Debor or Dabor, given to this wady by former authorities, may perhaps be ascribed.

About half a mile south of the pass, the Wady es Sidr receives an affluent which rises on the Mount of Olives, as Wady el Lehhâm. About two miles east of the Mount of Olives, the Lehhâm is joined by Wady el Haud from el 'Azîriyeh, and soon after the Wady el Jemel comes from Kh. Umm el Jemel on the south. The Wady el Haud takes its name from Ain el Haud, the "fountain of the Apostles," about a mile west of the junction, and the same has been applied generally to this wady, which continues from the confluence north-east to Arak; up to this place the Wady el Haud is

* Conder's "Tent Work," i, 299, 300, 301.

followed by the principal road between Jerusalem and Jericho. At Arak the Wady el Haud turns eastward to Wady es Sidr; and the bounding range of hills on the south takes the same direction towards Jebel Ekteif, from which it is only separated by the intersection of Wady es Sidr. At the south-western base of Jebel Ekteif, the Wady es Sidr enters the rocky gorge of el Mukelik, previously receiving two distinct wadys from the southern margin of the basin, named Wady el Mudowerah, and Wady ed Dekakin.

At Arak also, the road from Jerusalem forks in two directions. The road to the north-east goes to join the Ruabeh Road already mentioned. The road to the south-east goes over the hills to the junction of the Sidr with the Mudowerah, near the Mukelik gorge, and passes eastward along the brow of the range on the north of the gorge up to Neby Musa; whence it proceeds across the Ghor to join the road from Talat ed Dumm, at Ain Hajlah; whence it goes to the Pilgrims' Bathing Place at the Jordan. The branch road from Arak, by Neby Musa, is distinguished in the Survey as the Pilgrims' Road. From Neby Musa to Wady Sidr, this road was followed by Canon Tristram, on the way to Jerusalem, but the rest of the journey was performed across the country to 'Ain Haud. "Land of Israel," pp. 228, 229.

The Wady el Mudowerah has its most distant source at el 'Aziriyeh, which passes Abu Dis (alt. 2,100 feet) south-eastwards, and receiving several affluents from its neighbourhood, turns to the north-east and east at Wady Abu Hindi. On reaching Kabur el Madadi, the Wady Abu Hindi receives the Wady el 'Auwaj from the west, and after the junction it becomes Wady el Mudowerah, and pursues a very winding course to the Wady es Sidr.

The Wady ed Dekakin rises on the east of the hills where the Abu Hindi takes its north-east course, about two miles east of Abu Dis. As the Khaleil Abu Radt it runs to the east-south-east, till it receives an affluent descending in the opposite direction from the mountain on the waterparting named el Muntar (alt. 1,723 feet); then it turns to the north-east, takes the name of ed Dekakin, and receives several short tributaries from the waterparting on its way to the entrance of the gorge of Wady el Mukelik.

The rocky gorge of el Mukelik is about two miles long. It occurs where the basin contracts from a maximum width of more than six miles, to a minimum of a mile and a quarter. The Wady soon afterwards opens up into a great circular space formed by the slopes of Jebel el Kammun, Tubk el Kaneiterah, Jafet el Asla, and Neby Musa, and then passes at el Kueiserah to the plain, where it bends round south-eastward to the Dead Sea.

Besides the principal communications to which attention has been drawn, there are numerous tracks in all parts of the basin; indeed, Canon Tristram's excursion, before mentioned, proved that it might be traversed in some directions across country. One of these tracks, perhaps formerly of greater importance, appears to throw light upon King David's flight from Jerusalem by "the way of the wilderness," and finally to the Jordan and Gilead, 2 Sam. xv, xvi. Especially does it remove objections that have been made to the identification of Bahurim with Almon (Josh. xxi, 18) according to the Targum Jonathan. See Dr. George Grove's Art. "Bahurin" in Dr. Wm. Smith's Dict. Bib. Almon is also the Alemeth of 1 Chron. vi, 60, and the modern Kh. 'Almît, identified by Mr. Consul Finn and Dr. Tobler, and approved by Dr. Robinson ("Bib. Res." iii, 287), Van de Velde ("Memoirs," 284), Mons. V. Guerin ("Judée," iii, 75), and Lieutenant Conder. Monsieur Guerin's reference to 1 Chron. viii. 36, is curiously supported by the Survey, which besides representing Alemeth and Azmaveth, by the modern 'Almît and Hizmeh, adds also Wady Zimrij for the same personal name in this passage. Returning to Bahurim, as identified with Almon and 'Almît, it will be seen that there is a track passing from Jerusalem over the upper part of Brook Kidron and the northern part of the Mount of Olives, to Anâta, 'Almît, and Wady Fârah. Among the well-watered fastnesses of Wady Fârah, David may have taken temporary shelter, before he hurried away to the Jordan and Gilead. David's intention to take this route appears to have become known to Mephibosheth, who anticipated the entrance of David and his forces into the neighbourhood of his patrimonial estate around Jeba (Gibeah of Saul), by sending a present, which met David " a little past the top of the hill " (Olives) but which however the king refused to accept, in consequence of eliciting from Ziba, his master's intention to intrigue for the restoration of himself to the throne which his father had lost, and for which David had to struggle. Perhaps this rebuff became known to Shimei at Bahurim ('Almît), and excited the anger which he displayed along the hill side, without regard to his own safety, as David went either towards Wady Fârah or towards Hizmeh. The latter seems the preferable route, because, while David's immediate object was temporary shelter among the natural fastnesses by the way of the wilderness, his destination in the event of receiving unfavourable intelligence was Mahanaim on the north of Gilead. The temporary shelter which David selected with the further object in view, might have been the rocky fastness on the east of Jeba and south of Mukhmas, from which his friend Jonathan had expelled the Philistines. This natural fortress is three miles beyond 'Almît, and fifteen miles due west of the Jordan with a direct and main road to the fords. Here David got the advice to leave "this night and pass quickly over the water," 2 Sam. xvii, 16, 21, and he accomplished the march before daylight. If the more direct route to Gilead had been taken by Jisr Damieh and the River Jabbok, it could scarcely have been accomplished in a night by a considerable body of armed men, for the distance is not far short of eight-and-twenty miles. The shorter route is therefore preferred. The further discussion of this interesting period in the history of David, must await the extension of the Survey to the east of Jordan.

THE BASIN OF WADY EN NAR (BROOK KIDRON) AND THE MINOR BASINS ON THE NORTH.

The Minor Basins.

The outlet of the Wady en Nar into the Dead Sea, lies about two miles and a half on the south of the great headland of Râs Feshkah, and seven miles south of the outfall of el Kueisirah. This interval of seven miles is occupied by the following minor basins:—

(1.) Wady Jofet Zebeu rises between Tubk el Kaneiterah (alt. 306 feet, or 1,598 feet above the Dead Sea) and Jafet el Asla.

(2.) Wady Kumrân, called Maseb'a el Aîrneh in the plain bordering the Dead Sea, rises in the mountain of el Muntar (alt. 1,723 feet, or 3,015 above the Dead Sea), which stands between this basin and that of el Kueiserah on the north, and Wady en Nar on the south. The mountain is distant about eight miles from the outfall of Wady Kumrân in the Dead Sea. Wady Kumrân drains the extensive plateau of el Bukei'a, and a smaller but more elevated plateau on the west, called War ez Zeranik. El Bukei'a is about five miles long by two in width. It is traversed by a high road between Bethlehem, Mar Saba, and Jericho, which is joined at Mar Saba by a road from Jerusalem. Dr. Wilson came from Jericho to Mar Saba by this road, " Lands of the Bible," ii, 24, 25. Van de Velde crossed el Bukei'a from Mar Saba, " Memoirs," 117. See also Conder's " Tent Work," i, 298. Passes descend from el Bukei'a by the Wady Kaneitrah towards Jericho ; by Wady Kumrân to Kh. Kumrân, which M. de Sauley identifies with Gomorrah ; by Nukb Feshkah to Ain Feshkah ; and it has four or five communications with Mar Saba and Wady en Nar.

(3.) On the south of Wady Kumrân several short gullies intersect the lofty cliffs, but the most remarkable feature is the fountain of Ain Feshkah, which has been often described, Robinson's " Bib. Res." i, 533 ; Tristram's " Ld. of Is." 249.

(4.) Wady es Sammârah rises on the north of the Tubk es Sammârah and descends between it and Ras Muakit through a ravine on the south of Ras Feshkah. Near this wady and mountain is Kh. es Sumrah in the Plain of el Bukei'a. These with Wady 'Amrîyeh and the wady and ruin of Kumrân, in the same neighbourhood may also be reminiscences of Gomorrah. See Conder's "Tent Work," i, 298.

Wady en Nar Basin.

The northern boundary of this basin has been described in connection with Wady el Kueiserah from the north and east of Jerusalem to the summit of el Muntar. From that mountain to the north of its outfall at the Dead Sea, the boundary runs south-eastward to Khurbet Mird and skirting the southern end of el Bukei'a, ascends Tubk Sammârah, and descends to the southernmost foot of that mountain and to the Dead Sea.

The western boundary of the basin of Wady en Nar, begins on the Jerusalem road, about a mile on the south of Shafat, and takes a south-westerly direction to the Jaffa road, where the altitude is 2,669 feet, about a mile from the north-west corner of Jerusalem. Thence it pursues a south-westerly course to the Bethlehem road, having on the west the upper channels of the Wady el Werd in the Nahr Rubin basin. So far this basin impinges on the Mediterranean water-parting, but about half a mile north of the Kasr esh Sheikh, its western boundary, continuing on a south-easterly course, becomes divided from the Mediterranean basins by the Jordan basin of Wady ed Derajeh, which includes Bethlehem, and empties itself into the Dead Sea, seven miles south of the mouth of Wady en Nar, the interval being occupied by secondary basins. From the point of the divergence of this boundary from the Bethlehem road, it follows another ancient and straight road south-eastward to Kh. el Makhrum, where it is diverted to the south for a mile along the mountain of Umm et Tala (alt. 2,200 feet). Here the southern boundary may be said to begin. It takes up a south-easterly course along a range of hills to Maksar Ismaîn and Bir el Menwa, where it bends south for half a mile and goes on again eastward to the summit of Kurn el Hajr (alt. 1,460 feet). Here it turns to the north for half a mile, and then eastward to within a mile of the edge of the cliffs that overhang the Dead Sea at Ain el Ghûweir, and half a mile from the chasms that break down through them including Khashm el Hathrûrah. Here the waterparting bends round to the northward as far as Tubk Umm Keinis (alt. 617 feet or 1,909 feet above the Dead Sea), where it turns eastward to the outfall.

Watercourses of Wady en Nar Basin.

The valley of Jehoshaphat and the valley of Hinnom, east and west of Jerusalem are the heads of this basin. These unite in a deep valley at the Bir Eyûb or Joab's Well on the south-east of the city. The altitude of the valley at the junction is 1,979 feet, the hills on the north, east, and west, are respectively 2,518, 2,469, 2,549 feet.

From Joab's Well, the wady runs south-south-east towards Sheikh Sad, receiving a small branch from the valley between the Mount of Olives and Bethany (el 'Aziriyeh), and others of no importance at short intervals. Thence it proceeds almost south as far as its junction with the Wady Abu Aly, which rises near the waterparting and continues to skirt it for more than three miles. At the junction the Wady en Nar begins to bend round to the north-east and then slightly south of east up to the roots of el Muntar, when it turns south and enters the tremendous chasm of Mar Saba, which is about a mile and a half long between Bir Ibrahim and Bir ed Dikah, the convent being about midway.

Two main roads extend from Jerusalem to Mar Saba. One follows the western parting as far as Bir en Nefîs, before which it throws off two parallel roads eastward, the northern road following the summit of the hills that constitute the right bank of Wady en Nar, quite up to the convent; the southern comes from Bethlehem more than three miles to the west, and pursuing a more direct course across the heads of three wadys and the hills which separate them, joins the upper road about a mile from the convent. The other road follows the Wady en Nar, partly along the wady and partly on the left bank, up to a point facing Sheikh Sad, when the road continues its south-easterly course across the bend which the river makes in going to the south, and afterwards to the north-east. After coming again upon Wady en Nar, this road follows the stream up to the convent, generally parallel and within half a mile of the road on the top of the right bank. A track follows Wady en Nar round the bend, and others connect the different roads. The waterparting road from Jerusalem to Bir en Nefis, after throwing off the branches to Bethlehem and Mar Saba, pursues its south-easterly course first crossing the eastern head of Wady 'Alya, then following the ridge between the Wadys 'Alya and Surah, till it turns to the east by Wady el Abd, to the junction of Sirah and Theleithat; then over a stiff hill to Wady Umm Serj, which it descends to Wady Alya and Wady Jerfân. It leaves Jerfân where that wady turns to the north, and goes south-east over the waterparting, to cross the basin of Wady Ghuweir, and runs on south along the top of the mountainous crags which skirt the Dead Sea till it descends to Ain Jidy.

Below the chasm of Mar Saba, the Wady en Nar, turns eastward, till it is forced north-east and then south-east in cutting through the range that skirts the Dead Sea, between Tubk en Keinis (alt. 617 feet) and Tubk Sammârah (alt. 530 feet) to each of which must be added 1,292 feet for the depression of the Dead Sea.

The principal affluent on the left bank is Wady Akhsheibeh, which rises in the slopes of el Muntar, skirts the north-eastern boundary, and makes its junction with Wady en Nar, at the foot of Umm Keinis. About a mile above the last named junction the Wady Jerfân joins on the right bank. This Wady has its sources in a series of parallel valleys rising along the line of hills which cross the basin in a great arc from the western waterparting at Umm el

Tala eastward to Mar Saba. Of these Wadys (1) the 'Alya rises in Umm et Tala, skirts the southern waterparting for more than three miles, when it bends to the north-east, receives the parallel valleys and turns eastward, till it enters the Nusb Umm Seibeh, and comes out as Wady Jerfân, when it bends round to the northward to join Wady en Nar. (2) Wady el Areis, begins between Bir en Nefis and Kh. Deir Ibn 'Obeid, or Deir Dôsi, the remains of the very ancient convent of St. Theodosius (alt. 2,024 feet), from which Jerusalem is visible at a distance of five miles, Guerin, " Judée," iii, 88. The Wady Areis becomes Wady Surah and finally Wady el Abd, when it joins (3) Wady Umm eth Theleithât, which rises on the east of Deir Ibn Obeid. Their junctions together and with Alya are about half a mile apart. It is the heads of el Areis and eth Theleithat that are crossed by the lower road to Mar Saba from Bethlehem. The most easterly wady crossed by the lower road to Mar Saba, rises on the hill-side where the road bends southward towards the convent. Two of its valleys are on either side of the spur which is crowned by Burj el Hammâr, a ruined fort in a commanding situation. The valleys join at the end of the spur; and the wady, taking the names of el Makhrûm and Hajr, runs to Wady en Nar not far below the convent.

Such an outline of the hydrography of the famous Brook Kidron, as the foregoing, could not have been written before the new Survey supplied the materials for it. Some of its most important features have been hitherto quite misunderstood. The wady Abu Dis of Dr. Smith's Atlas, called by Van de Velde, Wady el Kazir, is the Wady Abu Hindi of the New Survey, and actually turns from south-east to north-west to join el Madowerah in the el Kueiserah basin. But in former maps it was continued into Wady Aksheibeh and so carried to Wady en Nar, out of its proper basin. The basin of Wady el Ghuweir, which is actually confined to the eastern slopes of Kurn el Hajr, within five miles of the Dead Sea, was extended back to the highland of Deir Ibn 'Obeid, and thus confused with the heads of Wady Jerfân which goes to Wady en Nar. The name of Deir Mirbeh, placed on former maps, between Deir Ibn 'Obeid and Mar Saba, can now only be referred to Burj el Hammâr, before mentioned. Perhaps the former name may be traced in the Bir Mukheibeh, a well at the foot of the Burj.

The Basin of Wady ed Derajeh.

The northern boundary of the basin has been described as far as Kurn el Hajr, where the interposition of the secondary basins of el Ghûweir and a series of gullies from the cliffs, causes the diversion of the present boundary southward, eastward and again southward to reach the Dead Sea.

The western boundary, starting from its northernmost point in the great plain (of Rephaim) on the south-east of Jerusalem,—continues along the main south road past Bethlehem, Urtas, and el Burak, to the Roman road which runs south from el Khudr, or the Convent of St. George. Up to this point the Wady ed Derajeh adjoins the Mediterranean basin of Nahr Rubin; but here it comes in contact with the basin of Nahr Sukereir, which runs by Ashdod to the sea. It now follows the road from el Khudr down to Kh. Beit Sawir, when it bends to the south-east to its termination at the head of the Wady el Biar.

The southern boundary proceeds from the head of Wady el Biar to the north by the Khurbet Breikût; then east to the Plain of Tekûa; where it turns again south-east to Kh. Tekûa (alt. 2,788 feet), and continues in the same direction for three miles; then it turns east to Bir 'Alla, then south-east to Kanan Rujm Kuddah, where it takes an easterly course along the edge of one of the rocky gorges of Wady Derajeh, until it is diverted southward along the eastern edge of Wady el Kurrât. From the conical hill at the southern end of that wady, it turns eastward to Ras Nukb Hamâr (alt. 678 feet or 1,970 feet above the Dead Sea), where it descends the lofty cliffs and crosses the plain to the Dead Sea, near the outfall of Wady ed Derajeh.

The Watercourses of Wady ed Derajeh.

Two main channels with a minor one on the south, divide the waters of this basin. These are Wady el Meshâsh, Wady ed Derajeh, and Wady Mukt'a el Jues.

(1.) The northern part has its heads as far south as the hills that divide Bethlehem from Urtas, the latter being in the southern division. The junction of all the wadys on the north of Bethlehem (alt. 2,530 feet) is on the road to Mar Saba, about two miles beyond the east end of the town. The northernmost is the Wady el Kaah, on the north of the village of Sûr Bâhir (alt. 2,612 feet). It descends to the junction along the northern edge of the basin. From a hill between Sûr Bâhir and Mar Elias, three main wadys run south to the Wady Samurah, which rises on the north of Bethlehem and runs eastward, receiving the valleys from the north, and then passing to the junction before mentioned.

Below the junction the wady is called Wady Lozeh, and runs to the southward, soon receiving the Wady Umm el Kulah from the south of Bethlehem, and the Wady et Tin from Beit T'amir. The Wady now becomes Wady el War, and flows south-east through a gorge, when it is named Wady D'abûb, and then Wady el T'âmireh. On reaching the southern base of Kurn el Hajr, it turns south and south-east as Wady el Meshash, and descends by a precipitous gorge to Wady ed Derajeh. Throughout this course the wady runs with the northern boundary of the basin, and receives short branches from it. In passing southwards from the base of Kurn el Hajr, it receives Wady el Bussah, and some smaller branches from the central range which divides the basin. This wady was formerly known in general as Wady T'âmireh, a name confined in the Survey to a small part of it.

(2.) The southern part has its origin in Wady el Bîar, which rises near Khurbet Breikût in the south-western angle of the basin, and skirts its western boundary northward up to Urtas.

[This vale was identified with the Valley of Berachah (2 Chron. xx, 26) by Dr. Grove in Smith's " Bib. Dict." ; but Lieut. Conder prefers Wady 'Arrub in the next basin (Handbook,

405); and Mons. Guerin argues in favour of a position on the south or south-east of Tekoa, and especially at Beni Naim, Judée, iii, 156. The situations of the "Wilderness of Jeruel," and "the end of the brook," are indicated hereafter on p. 242.]

Dr. Grove's opinion seems to be the most acceptable, because it is more on the direct road to Jerusalem, from the scene of the slaughter, which was on the way to Engedi; and it is also associated with an existing name, corresponding with the ancient one.

At Urtas the main wady bends round to the south-east to Kh. Bedd Falûn and Jebel Fureidis, ancient Herodium,* where it receives the outfall of several valleys rising on the east of Wady el Biar, and remarkable for the aqueduct which is carried along the hill-sides, zigzagging in and out of each valley in succession, on its way from Birket Kuffin near Beit Ummar to el Burak. Below this junction it is called Wady Fureidis and passes the ruins of Khureitun, and cliffs containing the great cave* which a false tradition identified with Adullam. The wady is now named after the ruined village, and runs on to meet the Wady Jubb Iblan and the main road which comes through it from Bethlehem to Ain Jidy. It is followed by this road south-eastward up to the entrance of the chasm of Wady Muallak, which becomes Derajeh lower down after its junction with Wady Mukta el Jass. The road turns to the south-west to avoid these great chasms of Wady Derajeh.

(3.) The Wady Mukta el Jass rises on the southern edge of the basin, at Khurbet Tekûa, the Tekoa of the Bible. It is called Wady el Menka, till it enters the chasm which leads it to Wady Derajeh, where it receives Wady Dannûn and Wady Bassâs.

THE BASIN OF WADY EL AREIJEH AND THE MINOR BASINS ON THE NORTH.

The next primary basin to Wady el Derajeh empties itself at 'Ain Jidy, the biblical Engedi (see a view in Tristram's

* Conder's "Tent Work," i, 294, 295. Guerin, "Judée," iii, 125-139 Robinson, i, 478-481.

"Land of Israel," 293), under the name of Wady el Areijeh, that name is however not even retained throughout the great chasm by which the main Wady descends from the upland to the sea, the upper part of the chasm being called Wady el Kelb.

The Minor Basins.

The distance between the outfalls of Derajeh and Areijeh is about seven miles. The interval is occupied by the secondary basins of Wady Hûsâsah, Wady esh Shukf, Wady Sideir, and a smaller one not named but delineated in the Survey, and called Wady Marjari in Canon Tristram's survey of the Dead Sea. See Map in Tristram's "Land of Israel."

Wady Hûsâsah (Robinson, "Bib. Res." i, 527) rises within four miles on the south of Kh. Tekûa, on the southern waterparting of the Derajeh basin; which, as it runs eastward through Bir 'Alla, and onwards as before described, forms the northern edge of the Hûsâsah basin. On the west the waterparting runs southward, and hugs the precipitous gorge of Wady el Jihar and its continuation as Wady el Ghar, as far as the straight ridge of Sahlet el Muhteirdeh, from which the boundary of this basin makes a sharp bend to the northeast, and again to the south-east, and again to the north-east, around the heads of the Wady el Mukeiberah, and so on to the cliffs at Abu el Rebâa, and the Dead Sea.

At the head of the Hûsâsah are four wadys, spread out over an area of three miles by two, and uniting at Bir Sukeiriyeh and el Megheidhat, where the altitude is 1,406 feet. These are surrounded by the waterparting on the north and west, and by an offset or spur from it on the south-east; which also throws off five branch wadys to the Hûsâsah, from its outer or south-eastern slope; the southernmost being also the recipient of branches derived from the greater part of the southern edge of the basin. The outlet of this upper plateau through the hills is below the junction of the wadys, where also three tracks meet from the north, south, and east; the passage westward being barred by the chasm of Wady el Jihar, in the next basin. The Hûsâsah makes bold sweeps to the north-east and south-east, in crossing its lower but still elevated plateau, to the chasm of its descent to the shore of the Dead Sea. The road between Engedi and Bethlehem, intersects the middle of the plateau obliquely, and the road between Engedi and Jericho crosses its eastern edge, at the back of the great cliffs which hang over the Dead Sea.

Wady esh Shukf is about four miles in length. On the north it has the south-eastern boundary of the Hûsâsah basin. On the south it is divided from the basin of Wady Sideir, by a range of hills extending from Ras esh Shukf (alt. 1,227 feet or 2,519 feet above the Dead Sea), to Khardet Hammameh. From this range the Wady Mekhowemeh descends south-eastward to the gorge of Wady Sideir, 'Ain Sideir, and 'Ain Jidy.

The Roads.

The following memorandum will explain the connection of these small basins with the roads of the country.

All the roads from Ain Jidi both northward and southward, ascend to the plateau by the same Pass called Nukb Ain Jidy, well described by Robinson, " Bib. Res." i, 501, 525. At the top of the Pass the roads divide northward and westward.

The western road runs along the edge of the gorge of Areijeh and Kelb, and is continued in the same direction by the track to Beni N'aîm, distant fifteen miles from 'Ain Jidy. At the third mile from the top of the Pass, this main road turns to the south-west, and crosses the gorge of the Kelb to the foot of the mountain called Kashm Sufra es Sana (alt. 1,400 feet, or 2,692 feet above the Dead Sea). Here it divides again, sending one branch in a curve south-west and northwest to Kurmul and Yutta, and another southward to Usdum.

The northern road skirts the edge of the Sideir gorge, and passes along the Wady el Mekhowemeh to the south-western base of Ras esh Shukf. Here it sends off to the north-west, the road to Bethlehem, which crosses the waterparting between the basins of esh Shukf and Hûsâsah at the Rujm Nueita, and descends by the Wadys Nueita and el Mukeiberah, to make its oblique passage across the Hûsâsah basin.

The northern road passes over the western side of Ras esh Shukf, with the hill of Jâfet ed Duwâ'arah further west, and crosses the Wady esh Shukf, to its northern waterparting at the head of Dr. Tristram's Wady Marjari. Further north it enters the Hûsâsah basin, crosses its gorge at Bir el Munkushîyeh and soon passes over the waterparting and descends into the gorges of Wady ed Derajeh.

The Basin of Wady el 'Areijeh.

The northern boundary has been traced in connection with the basins of ed Derajeh and Hûsâsah, up to the head of Wady Mukeiberah in the latter basin. Thence it proceeds southward, eastward, and south-eastward, along a range of hills dividing Wady el Kelb and 'Areijeh, from the small basins of Shukf and Sideir, and terminating at Nubk 'Ain Jidy.

The western boundary begins on the north, at a point on the main road between el Khudr and Beit Ummar, west of Khurbet Beit Sawir. It passes westward by Ballatet el Yerzeh (alt. 3,167 feet), towards the village of Safa where the waterparting follows a track running due south to Beit Ummar (alt. 3,010 feet).

It continues along the road till a track turns off on the south-west which it follows to Beit Sur, and then proceeds along a ridge to the high-road on the west of Hulhul, and terminates at Sîret el Bellâ'a (alt. 3,370 feet).

The western boundary of el 'Areijeh basin is conterminous with the heads of the valleys descending to Wady es Sunt in the northern part of the basin of Sukereir, which falls into the Mediterranean on the north of Ashdod.

The southern waterparting of el 'Areijeh, commencing at Sîret el Bellâ'a, passes eastward for about two miles, and then turns south-eastward by Kh. el Addeiseh and Bir el Jerâdât to the village of Beni N'aîm (alt. 3,120 feet). Hence it proceeds by a long ridge south-eastward between Wady el Kuryeh and Wady Umm el 'Aûsej to Kôd Ghanâîm, Hurubbet Umm el Kuleib, Khashm-Sufra-es-Sana, and the cliffs at the outfall of the wady into the Dead Sea. As far as Beni N'aim this boundary divides the basin from Wady el Khulil, which includes the city of Hebron, and runs by Beersheba to Wady Ghuzzeh, entering the Mediterranean Sea near Gaza.

The Watercourses of el 'Areijeh Basin.

A large and elevated plateau is enclosed at the head of the basin, between the waterparting and a long ridge extending from Kh. el 'Addeiseh to Bir ez Zaferân (alt. 3,000 feet), and Kanân ez Zaferân to the Wady el 'Arrûb, which divides the northern end of the ridge from a spur coming to meet it from the northern waterparting at Khurbet Tekûa (alt. 2,788 feet). It may be called the plateau of el 'Arrûb.

This plateau approaches the form of an equilateral triangle with sides of six or seven miles long. Its main wady and outlet is Wady el 'Arrûb, which rises in the north-western extremity of the basin, near Safa. It receives on its right bank all the wadys descending from the western edge of the basin, including—(1) Wady Marrina between Kh. Marrina and Kh. Beit Sh'âr; (2) Wady el 'Arab, and (4), Wady esh Sheikh, from Beit Ummar, Kh. Kuffin, and Beit Zata; (5) Wady esh Shinnâr, from Beit Sûr and the northern side of Hullûl; (6) Wady Siair, and (7) Wady ez Zaferân, from the south side of Hullûl, and the villages of Siair and esh Shûikh. On its left bank it receives the outlet of a group of wadys descending from an enclosure between the north waterparting and a spur from it following the left bank of Wady el 'Arrûb, on which Beit Fejjar (alt. 3,170 feet) is situated. These include (1), Wady er Rai, (2) Wady Bir el Khanzir, (3) Wady Beit Fejjar (separated by Rai and Khanzir from the village of that name), and (4) Wady Marah el Ajd. These fall into (5) Wady Rekeban, which empties itself at el Meniyeh into Wady el 'Arrûb, where that wady discharges itself from the plateau into the long and rocky gorge of Wady el Jihar and Wady el Ghar.

From the gorge of el Jihar and el Ghar, the main wady

runs on through three miles of open ground to the terminal gorge of Wady el Kelb and el 'Areijeh, all the way skirting the northern edge of the basin.

The Gorge of the Jihar and Ghar, with the three great branching gorges that fall into it from the west, belongs to a distinct plateau, bounded by the waterpartings on each side, also by the interior range of Kanan ez Zaferan, which divides this plateau from the more elevated plateau of Wady el 'Arrûb; and by another long interior range which proceeds from the waterparting at Beni N'aim and stretches across the basin eastward up to the left bank and lower end of the gorge of el.Ghar, where it tails off in a direction parallel with the waterparting at Sahlet el Mateirdeh. This latter range descends from the altitude of 3,120 feet at Beni N'aîm to 1,696 feet at Dahret el Meshrefeh. As the cross range on the upper side of the plateau maintains a height of not less than 3,000 feet, there is a cue to the depression of the lower plateau in the altitude at Dahret el Meshrefeh, which is supported by another on the same range near to it, at the well of that name. An altitude at the confluence of Wady Abu el Hamam with Wady el Ghar, or at the confluence of Wady el Jeradat with Wady el Jihar, would afford better indications. But unfortunately for geographers, it has not yet become the practice of scientific surveyors to apply their observations for altitude to the junctions of rivers and watercourses, as well as to the summits of hills and mountains; although of the two the hydrographic altitudes are of the more importance.

Between the cross range which forms the southern boundary of the Plateau of Jihar and Ghar, and the southern waterparting of the basin, there is a wady which rises on the north-east of Beni N'aîm as Wady Umm el 'Aûsej and after receiving Wady es Suweidiyeh, becomes Wady es Sûkiyeh, and falls into Wady el Ghar, before it drops down into the rocky gorge of Wady el Kelb. This wady together with the great terminal gorge, may be said to form the south-eastern division of the basin, or one of its three divisions which the present exposition of the new survey has brought to light.

A road crosses this valley and connects a track between Beni N'aim and Engedi with the road from Bethlehem to Engedi. It enters the basin at Kod Ghanaim on the southern waterparting, and goes across to Bir el Meshrufeh on its northern range, and then on to Bir Umm Jidy at the lower end of the gorge of el Ghar, and on northward across the Hûsâsah basin to the Bethlehem and Engedi road.

The Basin of Wady el Khubera.

The northern boundary of this basin, extending between Beni N'aîm and the Dead Sea, has been described. Its length is almost fifteen miles.

The Western boundary extends from Beni N'aîm southward, through Kh. Yukin, Tell ez Zif (alt. 2,832 feet), Kh. Ghanaim, el Kurmul (alt. 2,887 feet), and Kh. Maon, to Kh. Bir el Edd, a distance of about ten miles. It is throughout conterminous with the Wady el Khulil in the basin of Wady Ghuzzeh.

The southern boundary, beginning at Kh. Bir el Edd, runs north-eastward for three miles to Tell et Tûâny (alt. 2,637 feet). Then it turns eastward along the mountain of Dahret Hameideh to Khashm Sufra Lawundi and over the precipices to the Dead Sea.

The Watercourses of el Khubera Basin.

Above its gorge, the Khubera receives three important wadys, which are the outlets of three distinct parts of the basin. These are Wady el Jerfân, Wady Malâki, and Wady Rujm el Khulil.

I. The Wady Jerfân has its head wadys spread out along the western waterparting from Beni N'aîm to Kh. Istabûl. The northernmost branch comes from Beni N'aîm as Wady el Kuryeh, skirts the northern edge of the basin and descends to the Sahel or Plain of Abu el Ghuzeiyilât, where it receives the Wady Nimr and another wady from Kh. Yukin.

At the eastern end of the plain, the Wady es Sihanîyeh joins the Kuryeh. Its heads are found along the edge of the basin from Yukin to Istabûl. Two wadys descending from Istabul and Zif unite, and running eastward receive two branches from Yukin on the north, and one from el Bûeib on the south.

Below the plain the main Wady continues eastward along the northern waterparting as Wady Umm Kheiyirah and finally as Wady el Jerfân.

II. The Wady Malâki drains the rest of the western waterparting. Its principal source is in the south-west extremity of the basin, where it is first called Wady Kueiwis, and runs north-eastward to enter the gorge of Wady el War and Malâki. In doing so its name is changed three times in four miles. Before entering the gorge a branch is received on the left bank, which rises at the opposite extremities of a valley that runs along the waterparting between Kh. Ghanaim and el Kurmul. Another branch comes from Kh. Salma, following a ridge extending from Kh. Ghanaim to the entrance of the gorge. On the east of this ridge four valleys rise and unite before descending into the gorge on its north or left bank. The Malâki gorge is about six miles in length, and it is only separated by two miles of open ground from the Khubera gorge.

III. The Wady Rujm el Khulil rises on the southern margin of the basin at Khurbet et Tuâny, and drains a valley about nine miles in length between the margin and the Malâki gorge.

This basin is one of the scenes of David's exploits. Ziph, Carmel and Maon, and Engedi, retain their names unaltered except in spelling to this day. 1 Samuel, xxiii to xxvi. Lieut. Conder identifies the hill of Hachilah with the range on the north of the Malâki gorge, the summit of which now bears the name of el Kolah. The Sela or cliff of Ham-Mahlekoth (1 Samuel xxiii, 28) is referred to the Malâki gorge, still bearing the identical name. Conder's "Tent Work," ii, 89-91.

THE BASIN OF WADY SEIYAL AND THE MINOR BASINS ON THE NORTH.

Between the outfalls of Kuberah and Seiyal is an interval of five miles occupied by the minor basins of Wady Mahras, W. el Kasheibeh, and W. Sufeisif. The first two descend from the slopes of Khashm Sufra Lawundi, a summit rising at a distance of six miles from the Dead Sea.

Wady Sufeisif is only separated from the Mediterranean waterparting by the head of Wady Malâki. It rises on Tell et Tuany as Wady Meshukhkhan, runs eastward as Wady et Tebban and esh Sherki, and as Umm Merâdhif enters the gorge of Sufeisif.

Wady Seiyal.

The survey concludes with the left bank of this Wady. The western boundary begins with the ridge of Kanân el Aseif, and diverges to the south-east along the Khashm Beiyad which divides Wady Seiyal from W. el Kureitein in the Mediterranean Basin of Wady Ghuzzeh.

The northern boundary runs north-east with the head of W. Malâki for a mile and a-half; then turns to the south-east dividing the head of Sufeisif from W. Umm Jemat; then it makes a great curve to the north-east and south-east through Rujm el Bakarah and Khashm Umm es Suweid, follows the promontory between Seiyal and Sufeisif, and crosses the plain to the Dead Sea.

The Northern Watercourses of Seiyal Basin.

The main wady is represented as rising in Tell Arad, but the cessation of the survey leaves its development on the south quite unknown. On the north it soon receives Wady Khurbet et Teibeh from Kh. el Kureitein and the western waterparting, while the Wady el Kureitein, which was formerly supposed to join it, goes to Wady es Seba and the Mediterranean. The drainage of the north waterparting is collected by the Wady es Sennein, which joins the Seiyal in its gorge. The most western affluent of Wady es Sennein is the Wady Mutan Munjid, and the next is Wady Umm Jemat. The next descends through a rocky defile from Tawil el Butahiyeh and finally the Wady el Khuseibiyeh receives the rest of the northern drainage.

The delineation of the cliffs on the south of the outfall of the Seiyal has been considerately extended so far as to include Sebbeh, the site of the extraordinary fortress of ancient Masada, discovered by Dr. Robinson.* See p. 168.

Every one who appreciates the work accomplished by the Palestine Exploration Fund, must regret that its southern termination leaves so much beyond that is to a considerable extent utterly unknown, while altogether only scraps of information enable crude ideas to be formed of it. And yet it is a region in which some of the most interesting questions of biblical topography remain to be solved. It is the Negeb or South Country of the Scriptures; and the Rev. Edward Wilton's learned attempt to penetrate the darkness in which it remains, sufficiently expresses the scholarly interest which it attracts. No doubt the extension of the Survey in this direction will have the attention of the Managers of the Fund.

The division of the rivers into primary and secondary basins, has served to exhibit their relations to the great waterparting dividing the Jordan and Dead Sea from the Mediterranean Watershed or slope.

Attention has also been drawn to the parallelism which the upper valleys frequently bear to the great waterparting, which in respect to the general elevation of the country may be considered its main range. Such parallelism is a common feature of mountainous regions, and is perhaps most developed in the most mountainous. This remark is supported in a striking manner in Palestine by the occurrence of the most ample development of upper valleys parallel to the main

* Robinson, "Bib. Res.," i, 525, 526. Tristram's "Land of Israel," 303-314. Conder's "Tent Work," ii, 139.

range when the mountains are highest; as in the Litany along Lebanon, and the Wady el Khulil or Hebron Valley in the Mountains of Judea. The feature is one especially deserving of notice in correction of a common apprehension of the absènce of lateral communications in unknown mountains. The fair inference should be quite contrary.

PART III.
THE PLAINS OF WESTERN PALESTINE.

A further examination of the Survey points to aspects of the country which, while they are connected with its drainage, are not limited thereby. Such are the Plains and Highlands, which the Surveyors have delineated on a precise and accurate basis for the first time. A description of the lowlands, and of the most remarkable plains that are spread out among the heights of the interior, appears to be a fitting introduction to a study of the more complicated structure of the highlands which have the plains at their feet, and rise out of them. The famous Archduke Charles observed that " Once masters of the Plains, we are strategically masters of the Mountains," and the remark is as applicable to geography as to strategy. In this account the plains are arranged as follows :—

1. The maritime and upper plains on the Mediterranean slope.
2. The plains of the Jordan Valley.
3. The Western Shore of the Dead Sea.

1. THE PLAINS ON THE MEDITERRANEAN SLOPE.

The Plains of Galilee. The Maritime Plain of Tyre.

The northern extremity of the Phœnician Lowland, or the Plain of Tyre, is intersected by the River Kasimîyeh, at the northern limit of the survey. The southern end of the plain is at Ras el Abyad, the White Head or Cape, and the Promontorium Album of antiquity. This part of the maritime plain is about 13 miles long, and it was reputed to be " some three or four miles in breadth."* Kenrick made it five miles at Tyre.† The new survey has reduced these estimates considerably. Generally it is about a mile broad, and this is only exceeded very slightly near Tyre, while at each end of the section there is but half a mile between the hills and the shore.

* Robinson's " Phys. Geog. H. Land," 114.   † " Phœnicia," 19.

The most remarkable feature along the coast is the rocky promontory of Tyre, projected about a mile into the sea, and united to the mainland by an isthmus of half a mile in width, which affords shelter to shipping. An account of this famous site will appear in the "Memoirs," together with other particulars which cannot be noticed here. The plain is not very fertile, or its cultivation is limited, but there are large gardens for vegetables and fruit, besides patches of wheat and barley.

The wadys crossing the plain, beginning with the Kasimîyeh on the north, and ending with Wady Shema on the south, are notified in the first part of this introduction. The mountains on the east will be noticed hereafter.

From Ras el Abyad or the White Cape, to Ras en Nakura or the Hewn Headland, a distance of six miles, the mountains of Galilee extend either quite or near to the sea, and end in lofty white cliffs at the headland, and also at some intermediate places. They separate the Plain of Tyre from the Plain of Acre.

The Maritime Plain of Acre, and the Plain of Megiddo.

The northern limit of the Plain of Acre is the mountain range of Jebel Mushakkah, which terminates in the headland of Ras en Nakura. The southern limit is Mount Carmel, 20 miles distant. The coast runs in an unbroken line, a little west of south, as far as the fortified town and harbour of Acre, 'Akka, or St. Jean d'Acre, the biblical Accho and Ptolemais. Between Acre and the northern cape of Mount Carmel (Ras el Kerûm), the coast recedes back into a bay of eight miles in length, and varying in width from one mile on the north at Acre, to three miles on the south, where the port or anchorage of Haifa lies at the foot of Mount Carmel. Dr. Kitto described the bay as about three leagues wide, and two leagues in depth, but the meaning is not quite clear. From Acre to Jebel Mushakkah, or a distance of 12 miles, the plain is about four miles wide. It is intersected by the Wadys Kerkera, el Kurn, es Salik, Mefshûk, Majnuneh, and Semeiriyeh, which descend from the mountains of Upper

Galilee, and have been notified in Part I. The whole tract is fertile and well watered, and has extensive gardens and orchards, producing a variety of fruit and vegetables. Its general aspect is said to be that of a rich but neglected plain, where game of every kind abounds.

Between Acre and the foot of Carmel, the plain may be described in two parts, divided between the basins of the Nahr N'amein and the Nahr el Mukutt'a. Both parts have swamps near the coast; but the swamps of the N'amein are by far the more extensive, and have a length of not less than five miles, stretched out between Acre and the low hills of Shefa 'Amr. The plain is a noted pasture ground. In the northern part along the Wady el Halzûn, the Plain of Acre is more than eight miles wide, and generally five or six miles.

The hills that bound this part of the Plain of Acre on the east, form a range belonging to the western part of Lower Galilee. This range separates the Maritime Plain from a series of inland basins that are a characteristic feature of the region. The southern extremity of the range is defined by the Mukutt'a River, which separates it from Mount Carmel. From thence the range stretches in a north-easterly direction for about 22 miles, or as far as the eastern end of the Plain of Rameh, whence it extends towards the Plain of Gennesaret. It is the Northern Range of Lower Galilee. See p. 199.

The Plain of Rameh divides the hills of Lower Galilee, from a precipitous range that extends westward to Tell el Tantûr, opposite the city of Acre, and constitutes the natural southern termination of Upper Galilee. It rises to an altitude of 3,440 feet above the sea at Jebel Heider.

The further separation of the lower hills from the mountain range, is defined by a succession of wadys which connect the Plain of Rameh with the Plain of Acre, and form with the plain, a continuous passage for the high road between Acre and the Sea of Galilee, Safed, and Damascus. These wadys are—1. Wady esh Shaghûr which rises in the western part of the Plain of Rameh, and after running westward for about five miles, turns south to join Wady el Halzûn. 2. Next

Wady el Waziyeh, which overlaps Wady esh Shaghûr, and runs across the Plain of Acre, into the gardens near the city.

The Plain of Rameh runs up into the north-eastern extremity of the Nahr N'amein Basin, and somewhat beyond into the basin of Wady er Rubudiyeh, which falls into the Sea of Galilee. It is six or seven miles in length, and generally about a mile wide. Its altitude above the sea is about 1,200 feet. The central part is drained by a wady running south, which is met by another wady from the Plain of 'Arrâbeh, about three miles distant on the south-east. These wadys fall into Wady Shaib, which runs westward from the confluence, passing by the village of Shaib into the Plain of Acre.

The Plain of 'Arrâbeh, occupies the south-eastern projection of the Nahr N'amein Basin. It is three or four miles in length, and gradually expands to a width of two miles. Its altitude above the sea is perhaps 500 feet less than that of the Plain of Rameh, and probably does not much exceed 700 feet, but there are no heights recorded in either case. It has the villages of Deir Hanna, 'Arrâbeh, and Sukhnin on the surrounding hills. Two of these represent the Araba and Sogane of Josephus. Deir Hanna, although now only the ruin of a modern fortress, is an ancient site.

The plains to which attention will now be given, are beyond the range of hills that lie northward of Shefa 'Amr. The hills rise from the Plain of Acre, in a succession of wadys and rounded spurs, up to the summit of the range which runs north-east and south-west, and forms part of the waterparting between the basins of Nahr N'amein and Nahr el Mukutt'a. The altitude of the summit is 1,781 feet at two places, Jebel ed Deidebeh and Ras Hasweh. South of Shefa 'Amr, the waterparting bends to the north-west, and follows a spur of the hills to the Plain of Acre and the sea-coast. That portion of the hills which lies between this spur and Mount Carmel, is wholly in the basin of the Mukutt'a, and is cut through by the Wady el Melek, one of its principal affluents. But altogether the hills rise gradually to the waterparting from their base along the Plain of Acre and its connection with the Plain of Rameh; the slope running from about four to seven miles in length. The descent on the other side of the range faces the south-east, and instead of being long and gradual, it is short and sharp—a natural escarpment. The base of the hills on this side is in the great Plains of Buttauf and Merj Ibn Amir; the former having been the territory of Zebulun, and familiar to Josephus as the Plain of Asochis; the latter is the far more noted Plain of Megiddo, Jezreel, or Esdraelon, the chief battle-field of Palestine. These hills are part of the Northern Range of Lower Galilee. See p. 199.

The Plain of Buttauf, together with the smaller Plain of Toran, which lies at a higher elevation about a mile distant on the south,—occupies the north-eastern recess of the great basin of the Nahr el Mukutt'a or River Kishon of the Bible. Both plains are drained by wadys which are connected with the Mukutt'a by the Wady el Melek, which last joins the main river in the Plain of Acre. The hills which separate the Plain of Buttauf and the Plain of the Mukutt'a, rise but little above the plains, especially where the plains approach together most nearly; and remembering that both of the plains are parts of the same basin, it seems best to take a connected view of them, and to treat the Wady el Khalladiyeh, which is the outlet of Buttauf, as joined rather than separated by the low saddle which lies between it and the wadys on the south, that run down to the Mukutt'a, on either side of Kh. Zebdah (alt. 350 feet). Indeed Mount Carmel and the range of Samaritan Hills—which prolongs it in the same continuous line to the south-east beyond Jenin,—should be considered as the boundary of one and the same system of lowlands, in which the Plain of Acre as well as Megiddo, with their offsets would all be included.

Thus the remarkable features of these plains would be observed to the best advantage. Looking then from the point where the eastern face of the hills of Western Galilee meets the base of Mount Carmel, the plain stretches out its great arms to the north, north-east, east, and south-east.

Firstly comes the Maritime Plain of Acre, blocked on the north by Jebel Mushakkah, 25 miles distant.

Secondly, the offset between the slope of the western hills of Lower Galilee, and the lofty precipitous scarp that terminates Upper Galilee, beginning on the west with Wady Halzun and Wady el Waziyeh, and ending with the Plains of Rameh and 'Arrâbeh, being connected throughout as already explained.

Thirdly, the great north-eastern gulf or recess that runs along the eastward scarp of the western hills from Mount Carmel to Buttauf. This recess throws off three arms forming the head of El Buttauf, the Plain of Toran, and the broad depression between Seffurieh and Nazareth. These arms are divided by bold spurs projected from the water-parting of the Jordan Basin on the east. Jebel Toran divides the Plains of El Buttauf and Toran or Rummaneh, Jebel es Sik is the name applied to the eastern end of the spur which has Meshhed and Seffurieh on its summit, and it divides Toran from the low ground between Seffurieh and the hills of Nazareth. These Nazarene Hills complete the eastern boundary of the great north-eastern recess with its offsets, and divide it from the next great recess.

The Plain of Buttauf is between 400 and 500 feet above the sea, and the hills around rise to 1,700 feet. It is nine miles in length, and about two miles in breadth. The Jewish fortress of Jotapata, which was defended by Josephus against the Romans, is in a defile among the hills on the north. At the mouth of the defile is the ruin of Khurbet Kana, which Dr. Robinson claims to be the site of Cana of Galilee, where Our Lord celebrated a wedding by turning water into wine.* Rimmon of Zebulon (Josh. xix, 13) is identified with Rummaneh, at the entrance of the gorge leading to the Plain of Toran. The plain of Buttauf is one of the most fertile in Galilee. A great marsh occupies a considerable extent of its eastern part, and sometimes becomes a lagoon. Dr.

* Gospel of St. John ii; Robinson's "Biblical Researches," ii, 346-349; iii, 108. Smith's "Bib. Dict.," art. Cana.

Thomson found the track between Rummaneh and Kana flooded and dangerous.*

The Plain of Toran is 700 feet above the sea. It is about five miles in length, and seldom extends to a mile in width. On account of its fertility, it is called the Golden Plain. On its southern slope is the large village of Kefr Kenna, the traditional Cana of Galilee, and so regarded by the Latin and Greek Churches alike. The arguments in favour of the tradition are given at length by Mons. V. Guerin.† Lieutenant Conder proposes to remove the difficulty arising from the distance between Bethabara and Cana, by placing the former much nearer, and at the Ford of 'Abârah on the Jordan near Beisan, instead of at the fords of the lower Jordan near Jericho. But he is still disposed to prefer the traditional site for Cana of Galilee, before Dr. Robinson's; at the same time he points to another candidate in 'Ain Kâna, between Nazareth and er Reineh.‡

The mountain spur which bounds the Plain of Toran on the south, has the village of el Meshhed (alt. 1,254 feet) on its summit. It is the reputed birth and burial place of the Prophet Jonah.§ From this point the southern boundary of the Plain of el Buttauf, is found in the continuation from Meshhed of the hill which has Seffurieh on its summit (alt. 813 feet), and which descends to Wady el Khalladiyeh, where it meets a spur from Tell Seraj Allauneh, coming down on the eastern side of Wady el Ashert.

Between this boundary and the Nazarene Hills, which have their western termination at Semunieh (alt. 623 feet), is the hollow or low ground to which the name of the Sahel or Plain of Seffurieh may be conveniently applied and restricted towards the north; although it is at present written on the map somewhat further to the north than seems desirable for the purpose of geographical definition. The Kustul Seffurieh

---

* "Land and Book," 426.
† "Galilée" i, 168–182.
‡ Conder's "Handbook," 321; "Tent Work," ii, 64–68.
§ Guerin, "Galilée," i, 165–168.

comes out towards the middle of this plain, which forms the third and last offset from the great north-eastern recess. Seffurieh is all that remains of the ancient fortress of Sepphoris or Diocaesareia, anciently one of the strongest and largest places in Galilee, and often mentioned by Josephus, but never in the Bible. It is the reputed home of the parents of Mary, the mother of Our Lord. Semunieh is the poor remnant of the village of Simonias mentioned in Josephus and the Shimron of Joshua xi, 1 ; xix, 15.

The fourth great recess of the Plain of Megiddo lies between the mountain range of Nazareth and the isolated cone of Mount Tabor on the north, and Jebel Duhy or the Mountain of Neby Duhy, also Little Hermon, on the south. It is crossed by the waterparting between the Mukutt'a and the Jordan Basins, which descends from the summit of the Nazarene range about a mile east of Nazareth, and reaches the plain on the east of Iksal (biblical Chesulloth), passes to the middle of the plain on a south-easterly course, and at two miles from Iksal turns for about a mile to the south-west, and then southward to the foot of Jebel Duhy, ascending to the summit of the mountain by one of its shoulders on the east of the village of Nein, the Nain of the New Testament, where the widow's son was restored to life ; Luke vii, 11–18. Two miles from Nein, on the edge of the plain, is the village of Endor, retaining the name which it bore in the far-off days of Saul.

This portion of the waterparting divides the heads of Wady el Muweileh, an affluent of Nahr el Mukutt'a from those of Wady esh Sherrar, which becomes Wady el Bireh in its lower course to the Jordan. The receding plain begins between Tell Shadud on the Nazareth Road, and el Fuleh at the western foot of Jebel Duhy. It extends eastward for 12 miles, as far as the junction of Wady Shomer from the north, where the hills close in upon the descent of the valley to the Jordan. The length of the recess is equally divided between the two basins; but the width, which is from three to four miles up to Mount Tabor, becomes reduced to one or two miles on the south of that mountain. The Wady Bireh was traversed by Dr. Tristram,* but he was unable to visit the grand ruined fortress of Kaukab el Hauwa or the "Star in the Air," so called by the Arabs from its prominent and lofty site, at the junction of Wady Bireh with the Jordan Valley (alt. 975 feet, or about 1,900 feet above the Jordan). Descriptions of the view from Mount Tabor are given by Dr. Robinson;† and from Jebel Duhy by Lieutenant Conder.‡

The fifth recess is formed by the Valley of Jezreel or Nahr Jalûd, lying between Jebel Duhy and Mount Gilboa. It is rather a broad avenue to and from the plain, than a part of it, for at the entrance it begins a rapid descent from the plain to the Jordan at the average rate of nearly 80 feet to a mile. Still its general aspect persuaded Dr. Robinson to regard it as an arm or branch of the great plain.§ The waterparting from the summit of Jebel ed Duhy passes down to the plain at the village of El 'Afûleh in a south-westerly curve. The altitude at the village is 260 feet. From El 'Afûleh the waterparting curves round to the south-east as far as the village of Zerin (alt. 402 feet) at the northern extremity of Mount Gilboa; skirting in its course the edge of the descent of the valley of Jezreel. From Zerin the waterparting passes southward for a mile, and then turns south-east along the crest of Mount Gilboa, a name unknown to the natives, who have no general appellation for the range. Robinson describes the ascent to Zerin from the south as scarcely perceptible, but on reaching the village, it was found "standing on the brow of a very steep rocky descent of 100 feet or more towards the north-east."|| The valley has been already noticed in Part I.

The sixth and last recess of the great plain occupies the south-eastern angle of the basin, and lies between Mount Gilboa and the south-eastern extremity of the range which

* "Land of Israel," 453, 454. † "Bib. Res.," ii, 354.
‡ "Tent Work," i, 120. § "Bib. Res.," ii, 320.
|| "Bib. Res.," ii, 319.

continues along the plain in an unbroken line to Mount Carmel and the Bay of Acre. This range is formed by the final escarpment of the Hills of Samaria or Mount Ephraim. The entrance of this recess may be located along a track between Sîly and Zerin, where its width is six miles; and it runs back into the mountains for about nine miles, the width above Jenin and Arraneh being reduced to two miles.

The arms or recesses of the great plain are so thoroughly connected with the central part, that a separate account of each of them may appear to detract from the magnitude which the landscape in every case actually presents. Thus, from Jenin the view extends across the plain northward to Nazareth, a distance of 17 miles; although the imaginary line across the mouth of the recess is only six or seven miles away. Towards the north-west, the plain is seen extending for fully 20 miles, with Mount Carmel beyond. Similar remarks may be made in each case.

The central part of the Plain may be considered to extend along the Samaritan Hills from the gorge of the Mukutt'a, on the north-west, to Sîly on the south-east, a length of 15 miles. The width of this part may be taken between the Samaritan Hills and the ends of the opposite spurs which divide the arms and face these hills; and it may be reckoned at six or seven miles in each case, viz., the Nazarene Hills, Jebel Duhy, and Zerin at the foot of Mount Gilboa.

A few eminences occurring in this part scarcely serve to disturb its generally level aspect. The most remarkable is a low promontory thrown out from the Samaritan Hills towards the recess between the Nazarene range and Jebel Duhy. Its extremity is marked by ruins named Ludd (alt. 275 feet), the River Mukutt'a at the foot of the hill being 181 feet. On the other side of the Mukutt'a towards the north, there is an eminence of similar height, dominated by the hamlet of el Warakâny (alt. 277 feet). This appears to be connected with the more prominent termination of the Nazarene range at Ikneifis (alt. 508 feet). That a certain importance has been attached to these features is evident from the following

THE PLAINS OF ACRE AND MEGIDDO. 131

facts. The main road from the maritime plain of Sharon to Nazareth, passes from El Lejjun to Ludd, and then across the Mukutt'a, to what is probably a saddle connecting el Warakâny with Tell Shadud, near Ikneifis. On the south side of this pass, the Romans had the military station of Legio, now el Lejjun, and most authorities agree in placing the Jewish Megiddo among the remarkable ruins spread around the natural fastness of Tell el Mutasellim, on this promontory. This opinion is, however, not entertained by Lieutenant Conder.

The other principal roads in the great plain are : (1.) The highway from the coast at Haifa, along the foot of Carmel, and the continuing hills to Lejjun, Jenin, Beisan, and the east of Jordan. This road of course intercepts all communications from the north and south. A road from Acre falls into it at the gorge of the Mukutt'a, on the west of Sheikh Abreik, and it is followed by the telegraph line from the south. A road from Lejjun crosses the plain to the foot of Jebel Duhy, where it is met by roads between Jenin and Nazareth, both of which are centres of highways in all directions.

THE MARITIME PLAINS, SOUTH OF CARMEL.

The bold headland of Carmel, thrust out into the midst of the sea, appears to cut off communication along the shore, between the plain of Acre and the western side of the mountain. But it is not so, for unlike Ras en Nakura, the base of the mountain is separated from the coast line by an ample strand, traversed by one of the chief highways of the country. At Tell es Semakh, the width of the strand is 200 yards. Here is the northern end of the plain, that continues southward without much interruption throughout Palestine and the Desert beyond, to the shores of Egypt.

The width of the plain gradually increases, until about 9 miles south of Tell es Semakh,—at Athlit and beyond—the projection of the coast to the westward, extends the breadth of the plain to 2 miles. This continues with little variation for about 12 miles further south, when the Nahr ez Zerka, or

Crocodile River, crosses the plain, and the line of hills which has been followed so far for 21 miles, with descents often rocky and picturesque, comes to an end, in a bold bluff thrust out to the south, and filling up a great bend of the river. This feature is called el Khashm, and rises to a height of 554 feet, where it falls in with the main body of the highland. Several ancient sites are found along the narrow plain, which may be called the Plain of Tanturah. Tell es Semakh, at its northern end, is identified with Sycaminon. Ed Deir, with the Ashlul el Haiyeh, and the caves and springs adjoining, are the remains of the Convent of St. Margaret or St. Brocardus, the spring of Elijah, and the valley of the Martyrs (monks). Kh. Kefr es Samir, is the remnant of the Castra Samaritorum.* Athlit was the landing-place and castle of the Crusading Pilgrims, with Kh. Dustrey, or the Tower of Destroit, guarding a narrow pass in the neighbourhood. Jeba, is the "Geba of the Horsemen," colonised by Herod.†

Before leaving this part of the plain, it should be noticed that a line of sandhills, and sometimes rocks, skirts the shore as far as Athlit, and then continues on in the same direction, while the shore is advanced so as to leave a strip of land with jungle, between the sandhills and the sea up to Cæsarea, which is between two and three miles south of the Zerka. Between the ruined sites of Athlit and Cæsarea, are also found the ruins of Tantura or Dor, and Surafend, with some others.‡ Many Arab families find shelter among the ruins.

The Plain of Sharon.

The Nahr ez Zerka is the northern limit of the famous Plain of Sharon, which extends southward for 44 miles to the Nahr Rubin, and is there divided from the Plain of Philistia by the mouth of the river, and a line of heights on the south of Ramleh.

At its northern extremity, the Plain of Sharon expands eastward of the promontory of El Khashm in a bold recess,

* Conder's "Handbook," 210.
† Josephus, "Wars," III, iii, 11.
‡ Wilson's "Land of the Bible," ii, 248–253. Guerin, "Samarie," ii, 301–339. Conder's "Tent Work," ch. vi, vii; "Handbook," 310.

giving the plain a width of eight miles at Cæsarea. About six miles further south the width is ten miles, and towards Jaffa it extends to eleven and twelve miles.

The foot of the highland on the east is well defined all along the plain, for although the slope of the plain gradually approaches towards 200 feet above the sea, the hills rise sharply from it to 300 and 400 feet. Within four or five miles, altitudes exceeding 1,000 feet occur throughout the range, and further back the heights frequently rise above 3,000 feet. The foot of the highland is also generally marked either by the high road, or by another parallel to it.

The plain is by no means uniformly level, for blocks and ranges of low hills occur over a large extent of it. Some are well wooded, and there is quite a forest of oaks in the northern part near Kerkur. The low range along the coast which has been traced to Cæsarea, continues to Jaffa; and the line of hills in the narrow part of the plain north of Cæsarea, may also be traced southward to the same extent, in a much less elevated and less prominent shape.

Between Nahr Iskanderuneh and Nahr el 'Auja, there is a considerable cluster of hills about 20 miles in length, and from five to eight miles wide. Only their outer slopes contribute to the aforesaid rivers, which are here permanent. The interior constitutes a distinct basin drained by Nahr el Falik, which, previous to this survey, was supposed to extend far back into the mountains. The highest summit of these hills seems to be at Deir Asfin (alt. 302 feet), near which a saddle connects them with the eastern highland, and forms a portion of the waterparting between the basins of El 'Auja and Iskanderuneh. These Falik Hills are separated from the eastern highland by the Valley of Kulunsaweh on the north of the saddle, and by the Wady Kalkilieh on the south, and through both a high road passes from Jaffa to the north.

Another isolated mass occurs on the south of Nahr el 'Auja, and rises close to the gardens on the east of Jaffa. Its summit is named Dhahr Selmeh (alt. 275 feet), and it extends eastward for seven or eight miles, with traces of an ancient forest in that direction.

A more considerable group begins to rise between Jaffa and Ramleh; and the waterparting between Nahr el 'Auja and Nahr Rubin, crosses this range at altitudes of 240 and 260 feet, passing south of er Ramleh to Abu Shusheh (alt. 756 feet). This is the division between the Plain of Sharon and Philistia; the ancient Philistine border town of Ekron, being now found in the village of Akîr, about five miles south-west of er Ramleh. These may be called the Ramleh Hills.

The two principal groups of hilly ground divide the more level portions of the Plain of Sharon into three parts. The most extensive reaches from the northern extremity of the plain to the head of the Kulunsaweh Valley, a length of 24 miles. This tract is divided between three distinct river basins, which are those of the Nahr Zerka, the Nahr el Mefjir, and the Nahr Iskanderuneh. The waterparting on the plain between Zerka and el Mefjir is in the midst of the Oak Forest before noticed; that between Mefjir and Iskanderuneh lies between Kakon and Jelameh, and runs on to the semicircular line of hills with the village of Zelefeh (alt. 101 feet). The Zerka basin only runs back to the waterparting of the Mukutt'a between Jarah and Musmus. The Mefjir extends up to the Jordan basin on both sides of the Merj el Ghuruk. The Iskanderuneh basin is also in contact with the Jordan basin between Yasid and Mount Gerizim. (See Part I.)

The second and third portions of the level plain are in the basin of El 'Auja, and they are divided by the Dhahr Selmeh Hills. Through the central portion on the north of the hills, comes down the Wady Kanah and Wady Balut; the first by Jiljilieh, and the next on the south of Mejdel Yaba. These descend from the Jordan waterparting between Mount Gerizim and Bethel.

The southernmost level portion of Sharon receives the great Wadys Budrus and Selman. These barely touch the Jordan basin between Bethel and Bireh, being separated from other parts on the north and south, by the overlapping of the adjoining basins. This division of the 'Auja basin is of equal width with the northern division. The Budrus enters the plain on the north of Haditheh, as Wady es Surar. The

Selman passes on the northern side of the town of Ludd, as Wady Ludd and Wady Razia, and it unites with the Budrus at Kefr Ana in the midst of the level.

The view of all this region from the Tower of Ramleh is described by Dr. Robinson as "rarely surpassed in richness and beauty."* Dr. Thomson declares it to be "inexpressibly grand, the whole plain of Sharon from the mountains of Judea and Samaria to the sea, and from the foot of Carmel to the sandy deserts of Philistia, lies spread out like an illuminated map."†

Such are the features of this famous plain as they have been for the first time clearly defined by the survey. Its villages and ancient sites will be noted in the Memoirs. The most complete account of them, hitherto, is given by M. Guerin in his volumes on Samaria, but he was unable to detect the erroneous representation of the principal streams which was then accepted, and thus confounds Wady Shair with Nahr el Falik, with the result of confusing Wady Shair with Wady Kanah, which is 15 miles further south, and also Micmethah with Kakon.‡

The following upland plains are connected with the maritime region in various ways, which will be explained.

The Plain of 'Arrâbeh or Dothan.

The Nahr el Mefjir crosses the Plain of Sharon on the south of Cæsarea, and the upper part of its basin includes a series of plains which, but for the outlet of their waters into the Plain of Sharon, would be more nearly related to the south-eastern recess of the Merj Ibn Amîr, the Plain of Esdraelon or Megiddo.

The Plain of 'Arrâbeh or Dothan, where Joseph was seized, and sold by his jealous brethren, is connected with the maritime plain by Wady Abu Nar, and continues to be a highway between the Mediterranean and the east of Jordan, as it was in Jacob's time. Dr. Robinson, on his last journey,

---

* "Biblical Researches," ii, 231.
† "Land and Book," 530.
‡ Guerin, "Samarie," ii, 346.

viewed the plain from Yabud, a village on the hills above its western end; and he noticed its outlet through Wady el Wesa, which lower down becomes Wady el Ghamik, and afterwards Wady Abu Nar, an affluent of Nahr el Mefjir.*

The distance between the plains of Sharon and 'Arrâbeh is eight or nine miles; and between the eastern end of the Plain of 'Arrâbeh and the Plain of Megiddo, is only two miles. Dr. Robinson describes the Plain of 'Arrâbeh as "appearing like a bay or offset, running up (from the Plain of Megiddo), among the southern hills."† Altogether the distance between Jenin and the Plain of Sharon is 17 or 18 miles. Further north, the distance between the plains of Sharon and Megiddo is about 12 miles, following the high road between Kh. es Sumrah and el Lejjûn. The altitude of the plain above the sea is about 800 feet at its eastern end, and 700 feet at the western outlet.

At its south-eastern extremity, the Plain of 'Arrâbeh is connected with a series of upland plains extending from the southern end of Mount Gilboa, along the Mediterranean and Jordan waterparting to the mountain range on the north of Samaria; these are:—(1.) The upper plateau of Wady es Selhab. (2.) The Merj el Ghuruk, devoid of outlet, a luxuriant corn-field in summer, and a marsh or lagoon in winter. (3.) The plain on the north of Fendakumieh noticed by Van de Velde ("Memoirs," 236). The altitude of these three plains is about the same, viz., 1,200 feet.

The Plains of Mukhnah, Rujib, 'Askar, and Sâlim.

The northern end of these continuous plains lies on the east of Nablûs, from whence they extend in two arms, one of which runs for six miles southwards along the Jerusalem road; while the other stretches for five miles to the south-east, and terminates at Tana. The southern part, or Sahel Mukhnah, is drained into the Plain of Sharon, 17 miles distant, by the Wady Kanah; but there is no other connection with the maritime plain except by mountainous tracts. The other parts have their

* "Bib. Res.," iii, 123, 124.
† Rob. "Phys. Geog. Holy Land," 122.

waters carried to the Jordan by Wady Fâr'ah (see page 73). The altitude varies from 1,800 feet at the head of the eastern arm, to 1,600 feet at the outlet of the southern end.

The Plain of Philistia.

From the sandy range of hills on the south of Ramleh, the plain extends to Gaza and the extremity of the survey; beyond which it is prolonged through the desert called in Scripture the Wilderness of Shur. The length of the Philistian coast-line is about 40 miles; from Ekron on the north to the hills on the south, which divide Wady Hesy from Wady Sheriah, is only 30 miles. The difference is mainly due to the obliquity of the coast-line.

The whole of Philistia is included in the Shephelah, or lowland country; but like the lowlands of Scotland, the plains are only a part of the lowland, the hills forming a large proportion of the total area. The eastern border of Philistia or the Shephelah, dividing the lowland hills from the Highland of Judæa, can now be traced through the natural features of the country made known by the Survey. For the present the plain alone will be discussed.

Like the Plain of Sharon, this southern tract is also broken up by low hills, which are quite inferior to the more elevated division of the lowland. The long line formed by the descent of the highland to the Plain of Sharon, is in some degree preserved in the Philistine plain, as far as 'Arâk el Menshîyeh; but the whole aspect is less regular. At Akir (Ekron), and at 'Arâk el Menshîyeh, there are broad plains, which make recesses in the general line of the higher hills. These are watered or intersected by the main branches of the Nahr Rubin and the Nahr Sukereir respectively.

Between these plains are low hills rising to 300 feet, and enclosing a hollow tract about five miles across, with the village of el Mesmîyeh in the centre, and others around, drained by affluents of the Sukereir. Through the south of this hollow passes the Wady es Sunt, which, under various names, collects many branches from the highland between Bethlehem and Hebron, and brings them into a focus at the foot of the highland in the Valley of Elah, the battlefield of David and Goliath. From hence the wady passes westward through the hills of the Shephelah by a crooked gorge, which begins on the east at the foot of Shuweikeh, the site of Shocoh, and debouches into the plain at the foot of Tell es Safi, the Crusaders' Blanchegarde, and most probably the biblical Libnah. The plain here consists of the low rolling plateau which encloses the hollow around Mesmîyeh. Over this plateau fled the Philistines after the death of Goliath, northward to the Plain and City of Ekron, and southward across the Plain of 'Arâk el Menshîyeh to Gath, by the way to Shaaraim. 1 Sam. xvii, 52.

The Nahr Rubin comes down to the Plain of Ekron from sources at Beeroth, which lies due east of the outfall; but the basin makes a great semicircular sweep to the south, including the Plain of Gibeon, and passing west of Jerusalem; it also takes in the Plain of Rephaim, and extends as far south as 'Ain Shems (Beth Shemesh), where it is at the foot of the highland, having collected all the drainage between Beeroth and Bethlehem. From 'Ain Shems it passes through the hills of the Shephelah, by the broad Wady es Surar, in the midst of Samson's country, which looks down on the Plain of Ekron. The plain is about six miles across, and a spur from the Ramleh heights separates it from another on the coast, about four miles in extent, and containing Yebnah (Jabneel), with other villages. Between the Plain of Yebnah and the sea-shore, is a sandy down three miles in width, and forming part of a series, which from Jaffa southwards, takes the place of the narrow sandhills which skirt the shore northward.

The Plain of 'Arâk el Menshîyeh (so called here for convenience in the absence of any other name), is more considerable than the Plain of Ekron, being 14 miles in length from the eastern hills to Esdûd (Ashdod), and four or five miles in breadth. Dr. Robinson crossed from Tell es Safi to Keratiya on the way to Gaza, and found the scene enlivened by large herds of cattle and flocks of sheep and goats; —the country beautiful and fertile, almost perfectly level,— with a light-brown loamy soil. The crops were good, yet hardly half of the plain was under cultivation (Robinson's "Bib. Res.," ii, 32).

The wady which enters the plain at Zeita (alt. 518 feet), comes down from the highlands between Hulhul and Hebron, where the culminating summit has an altitude of 3,370 feet; the direct distance from Zeita being only 17 miles. The plain below Zeita is probably about 300 feet. Ashdod, on a hillock (alt. 140 feet), at the western end of the plain, is now separated from all that remains of its port, by sand-downs, three miles in breadth. The site is occupied by the present village of Esdûd, with 1,800 people, but the remains of this primæval city, once so strong and mighty, are so few and insignificant, that one is tempted to suppose the greater part of the city may be buried beneath the sands. If so they may be in a superior state of preservation, and perhaps repay for exhumation.

From Khurbet Yasin on the south of Esdûd, or Ashdod, to 'Arâk el Menshîyeh, a range of hills running in a southeasterly direction, forms the waterparting between Sukereir and el Hesy basins, and bounds the plain on this side, increasing in height as it proceeds inland. Beginning with 120 feet at Kh. Yasin, the altitude has become in seven miles 331 feet at Kh. Ejjis er Ras ; and the hills so far divide the basins of Sukereir and el Bireh. Beyond this point the range divides the basins of Sukereir and el Hesy. At Tell Ibdis, seven miles farther, the altitude of the range has increased to 450 feet; and nine miles beyond, at Sheikh 'Aly, in the higher part of the Shephelah, it is 1,367 feet. It may be called the Sukereir Range.

South of this range, the aspect of Philistia undergoes a change. The plains give way to hill and dale. While heights of 1,000 feet above the sea maintain a regular alignment almost due south, from Surah on the north in Samson's country, all the way to Kuweilfeh towards Beersheba in the south, the lower slopes beyond the range are advanced westward, in correspondence with the advance of the shore line in that direction. North of the range, the foot of the hills is found as far east as Zeita and Tell es Safi; although here exceptional indications of the advance occur at Berkusieh and Summeil. South of the range the advance becomes general, and is defined by the fine valley which crosses the range from Keratiya in the plain, to Bureir, Simsim, and Tumrah, on the way to Gaza. The change is modified here also, by the valleys which run back from Bureir into the hills, at Tell el Hesy and Huj. South of Tumrah the mass of the hills recedes eastward, and nine miles south-east of that place they send a spur to the south-west, dividing the broken plain on the east of Gaza from the pastures of Wady esh Sheriah.

The range of hills on the west of the Bureir valley is also more elevated and bolder than any in a similar position further north. It has a culminating height of 426 feet, and extends from Esdûd to Simsim and Deir Sineid, on the north of Gaza. Of course these hills are traversed in all directions; but the main road from Gaza to the north follows their western side, and sometimes encroaches on the sandy downs towards the sea.

North of Gaza, the sand-downs are interrupted by the passage to the sea of Wady el Hesy, with the village and gardens of Herbieh on its north or right bank. Mons. V. Guerin gives the Wady el Hesy in this part the name of Nahr Eribiah or Nahr A'skoulan, evidently from the village of Eribiah or Herbieh, and the ruins of ancient Askelon on the north. But it should be observed that Askelon is not included either in the basin of Wady el Hesy or in the partly adjoining basin of Nahr Sukereir on the north. It is in a small coastal basin (like that of Nahr Falik beyond Jaffa) which is chiefly drained by Wady el Bireh and its affluents, and has its outlet in a sink among the sand-downs on the north of Hamameh. The former outlet of this wady, before the sands blocked up its channel, seems not unlikely to have been at Bir esh Shekeir, where there is an opening in the cliffs which line the coast. In this direction also, may have fallen the drainage of Wady el Jabbar near el Mejdel, and· that of the valley on the west of El Mejdel and Hamameh. If the outfall of the basin was at Bir esh

Shekeir, then it is there that the remains may be found of Maiumas Ascalonis, or the Port of Askelon, which Mons. Guerin sought for in vain along the coast further south, sunset having prevented him from extending his search for the port, northward of Skeikh 'Awed (Oualy ech-Cheikh Haoued) Guerin, " Judée," ii, 151, 152.

The maritime belt extending from Gaza and Askelon, back to the hills bounded by the Valley of Bureir, is under the control, exercised from the earliest ages, of one or the other of those ancient places. The whole of Northern Philistia with its open plains, felt the weight of Ashdod up to the crusading times. To guard the northern frontier towards the interior, there was Ekron, protecting the passes by Ramleh and Wady Surar. The southern frontier towards the interior appears to have been similarly guarded by Gath. These were the five dominant cities of ancient Philistia. The sites of four of them are well known, only Gath is still in dispute.

The new survey now shows not only the names of ancient sites, but also the natural features of the ground; and thus the probable situation of a place intended to guard the inland part of the southern frontier of the country, may be examined irrespectively of other considerations. Wherever the hills on the east of Gaza afford the readiest access from the south into the Philistine Plains, there Gath should be sought, especially in connection with the " Way to Sharaaim." 1 Sam. xvii, 52.

An examination of the hills of South-Eastern Philistia shows that the passage from the northern plains to Wady Sheriah, so as to keep out of the way of Gaza, is most easily made along a line running southward from 'Arâk el Menshiyeh in the south-eastern angle of the plains. The tendency of the hills and valleys is here north and south. The ground also forms a sort of terrace or plateau between the descending slopes westward and the ascending slopes eastward. Westward of this line, the valleys and ranges run towards the coast, and form a succession of "ridge and furrow" on too obstructive a scale for a through route across them. Eastward of the same line, the country rises rapidly from a mean altitude of 500 feet, to heights exceeding 1,000 feet; and the routes in the more elevated region would lead into the poor hills instead of the rich plains. The salient points in the higher range are :— Sheikh Aly (alt. 1,367 feet); Reshm esh Shukkâk (alt. 1,085 feet); Kh. Umm Kushram (alt. 1020 feet); Kh. esh Shelendi (alt. 1,130 feet).

Such is the condition of the ground connected with the investigation of the site of Gath, to which attention has been more especially given in the Palestine Exploration Fund *Quarterly Statement* for October, 1880.

2. THE PLAINS, LOWLANDS, AND LAKES OF THE JORDAN SLOPE.

In the examination of the Plains and Lowlands of the eastern side of the Highland of Western Palestine, it should be borne in mind that this division of the subject can only be treated partially, in consequence of the survey being terminated on the east partly by the River Jordan, and partly by the western shores of its expansions into lakes. But obviously, the features now coming under consideration embrace both sides of the water whether those features be plains or gorges or inland seas. It may therefore be desirable now and then to take the eastern side into view, on the basis of the information in existence beyond the survey. But after the misconceptions and inaccuracies brought to light on the west by the survey, it would be rash to venture far into the obscurity which lies beyond it.

At the commencement of Part II. a few remarks have been made upon the features of the Jordan basin at its origin, in explanation of these occurring at the northern boundary of the survey including the Merj 'Ayûn, which is the northernmost of the plains on this side. The following notes on the Merj 'Ayûn will be in continuation of those then made in connection with it.

THE PLAIN OF MERJ 'AYÛN, OR THE MEADOW OF THE FOUNTAINS.

Only the southern part of Merj 'Ayûn falls within the survey. Its name Ayûn, according to Dr. Robinson, is the equivalent, in Arabic, for the Hebrew Ijon of the Bible (1 Kings xv, 20; 2 Kings xv, 29; 2 Chron. xvi, 4), the meaning of both being the same. The ancient city of Ijon is identified with Tell Dibbin, a mound beyond the map, rising 110 feet above the base; a noble site, overlooking the whole plain, and commanding the great road from Sidon to Damascus.* The map begins at 'Ain Derderah in the centre of the plain, which is said to be six miles long, and from one to three miles broad. It is connected with the Huleh Plain and the Jordan by Nahr Bareighit, which unites with the Hasbany, near the junction of the Leddan and Banias.

Dr. Thomson describes a furious storm which occurred in this plain in December 1856. It came on with great rapidity in columns of mist from the Huleh. Ten men in full view of their homes were unable to escape from it, and died in a few minutes from its intensely chilling effects. There was neither snow nor frost nor much rain, but the force of the wind tore up and drove everything before it. Eighty-five head of cattle perished at the same time, chilled to death by the wind.†

An aneroid observation by Dr. de Forest, quoted by Dr. Robinson, gives the northern end of the plain an altitude of 1,822 feet, the southern end being 1,500 feet, but the slope is imperceptible, the surface appearing to be quite level. The French Survey of the Lebanon, according to the "Carte du Liban," gives the altitude of Tell Dibbin as 1,770 feet, which is confirmatory. It has been said that the natural course of the Litany River is southward to the Jordan Valley,

* Robinson, "Bib. Res.," iii, 393-375. Guerin, "Galilée," ii, 280. Thomson, "Land and Book," 222-225.
† "Land and Book," 224.

that is, through the Merj Ayûn; and volcanic operations are presumed to be the cause of the formation of the ground which has diverted the river to the west.* It is therefore desirable to compare the height of the plain with that of the River Litany or Kasimîyeh, and with that of the Jordan in the Huleh Plain. The height of the Litany at Jisr Kardeli, the bridge below the Merj Ayûn, is, according to Dr. de Forest, 700 feet. The waterparting between the River Litany and Merj 'Ayûn is not observed, but farther south at Neby 'Aueidah, the Palestine Exploration Survey gives 2,814 feet. The range which divides Merj 'Ayûn from the Hasbany River at Abl is 1,704 feet. North of Merj Ayûn, where a single ridge divides the Litany from the Hasbany, between Kaukaba and the Burghuz bridge, the altitude is 2,300 feet. The altitude of the Huleh Plain is 140 feet, at the junction of all the streams which the Jordan carries into the Huleh Lake. A few more altitudes between the elbow of the Kasmîyeh and the Huleh are much wanted. For altitudes anterior to the Palestine Exploration Survey, *see* Van de Velde's "Memoir," 1858, and " Notes," 1865.

The Huleh Plain, Marsh, and Lake.

The total length of this low region is about 16 miles, and the width six miles. It is naturally divided into four distinct parts, namely (1) The Huleh Plain; (2) The marshes; (3) The Plain or Ard el Kheita; (4) The Bahr, Baheiret, or Lake of Huleh.

The Huleh Plain extends from north to south between four and five miles, or about one-fourth of the total length. It is bounded on the west by the mountain range which divides the Jordan from the Kasimiyeh, including Jebel Hunin, with an altitude of 2,951 feet, at the north-west angle of the plain. On the east it has the south-western roots of Jebel esh Sheikh or Mount Hermon, with an altitude at the Castle of Banias or Kulat Subeibeh of 2,485 feet. The northern boundary consists of high ground between the eastern and western mountains, rising to heights of about 1,000 feet, and

* " Palestine Exploration Survey Memoirs," Sheet ii.

descending by slopes and terraces to the plain. Dr. Robinson's account of this descent is very explicit, but the Survey, while partially confirming the detail, does not fully elucidate it.*

The last step or offset to the lowest plain is placed by Robinson at el Mansury (Mansûrah), where the Survey denotes the elevation by the altitude of 245 feet, but does not mark the edge of the step. See p. 181.

The southern limit of the Huleh Plain is the great marsh, which stretches across from the eastern to the western hills, in an unbroken line. Where the Jordan enters the marsh the mud village of es Salihîyeh is situated. Altitudes are wanting.

The plain is exuberantly fertile, and considering its contact with the great marsh, it is worthy of note that the plain is free from marshy ground in the dry season; but in wet weather the ground is widely flooded.

The Huleh Marsh.

The marsh extends from es Salihîyeh in the plain, to the northern edge of the lake, a distance of nearly six miles. There is a variable space between it and the foot of the western hills, amounting at the most to three-quarters of a mile, along which a track passes. The eastern side is excluded from the Survey, but according to Dr. Tristram ("Land of Israel," 585), the plain becomes swampy up to the ford of the River Banias, which seems to be at el 'Absîyeh, two miles and a half north of es Salihîyeh. Sukeit, which he reached after three hours' riding, does not appear to be a mile further south than es Salihîyeh, and therefore it becomes difficult to account for his "floundering through several miles over swampy plain," and miles more through ripe wheat and cotton plants. The marsh seems to extend to the foot of the eastern hills. Dr. de Forest passed along the upper part of these hills, but no traveller seems to have attempted to trace the foot of them. Mr. J. Macgregor, in the "Rob Roy" Canoe penetrated the only open channels that he was able to find among the reeds both on the north and south sides of the

* "Phys. Geog., H. Land," 68 ; "Bib. Res.," iii, 389.

marsh, but no through passage could be found; and the reeds enclosing these channels resisted every attempt to force a way through them. A continuous channel is delineated in the New Survey, but no explanation of it has yet appeared.

The edge of the marsh for half a mile is a belt of ordinary bog, up to the knees in water. Then comes a deeper belt where yellow water-lilies flourish. Then a belt of tall reeds with white water-lily in the open spaces. Beyond is an impenetrable wilderness of Papyrus or Babeer extending to the eastern side. This is a thin floating crust of vegetation over depths of about twelve feet of peaty mud and water. The only footing is on the slippery roots of the papyrus. Both Dr. Thomson and Dr. Tristram nearly lost their lives among it in pursuit of wild fowl.*

The Huleh Lake.

The lake is triangular, with its apex at the south, where it runs into the Jordan. It commences at the southern end of the marsh, and is about four miles long from north to south, its breadth along the marsh being perhaps greater, but the eastern part is unsurveyed. Mr. Macgregor states that "the lake lies quite close to the hills on the Bashan side." "Rob Roy on the Jordan," 6th edit., p. 271. But the hills have never been properly delineated, nor has the practicability of passing along the eastern edge of the water been yet tested. Dr. Tristram describes the western edge as "fringed for the most part by a bank about six feet high, below which is a narrow strip of deep shingle formed chiefly of the *débris* of shells, and the bank waving with wheat to its very edge. The lake had been five feet deeper in winter, and its ordinary height might be told by the fringe of oleanders, which grow stilted like mangroves with several feet of root at present high in the air. The water was shallow at this side, for acres of yellow water lilies floated on the surface, and a few patches of white nymphæa grew behind papyrus tufts." "Land of Israel," 589.

The observations of the Palestine Exploration Survey, make

---

* Thomson's "Land and Book," 257. Tristram's "Land of Israel," 588.

the altitude of the surface of the lake seven feet above the sea, which gives a fall of 133 feet from the confluence of the rivers in the Huleh Plain. Captain Mansell, R.N., made it 273 feet, and De Bertou reported 20 feet below the sea level. Other erroneous attempts are quoted in Van de Velde's Memoirs. The only soundings upon the lake were made by Mr. Macgregor in "Rob Roy," and were found to vary from 9 to 15 feet.

The Ard (Plain) el Kheit.

The beautiful and fertile Plain of el Kheit begins on the north at 'Ain el Mellâhah and terminates on the south at Kh. el Muntar, which is on a range of hills forming the waterparting of small streams on the south that flow into the Sea of Galilee. The lower part of the plain skirting Huleh Lake is perfectly level; but towards the mountains on the west, it is diversified by rolling ground. The length of plain is about eight miles, and its average breadth about four. The high road which crosses the Jordan at the Jisr Benât' Y'akûb (Bridge of Jacob's daughters), skirts the southern edge of the plain; and it is traversed by tracks along the foot of the mountains, also along the lake, and in many other directions.

The northern part of el Kheit is watered by brooks which fall into a stream rising in the fine fountain of 'Ain el Mellâhah. The southern part is crossed by Wady Hindaj which descends from Jebel Jermuk (alt. 3,934 feet) through deep and rocky chasms. Wady el Wakkas follows. It rises in the Merj el Jish on the north of Safed, and breaks its way down to the plain by a precipitous course. Another wady crosses the plain from Kh. el Loziyeh, and joins the lake near its outfall.

THE GALILEAN GORGE OF THE JORDAN; OR THE DESCENT FROM LAKE HULEH TO THE SEA OF GALILEE.

The waters of Lake Huleh are nearly on a level with the sea. About six miles and a half north of the lake where the Jordan unites its affluents before entering the Huleh marsh, the altitude is 140 feet above the sea. About a mile and a half after the river has left the lake, and near the Bridge of Jacob's daughters, the river has already fallen 43 feet below the sea level. About eight miles further south the river has descended to the level of the Sea of Galilee, or 628 feet below the Mediterranean. This long and rapid descent takes place in a deep gorge, and forms a continuous cataract without any prominent cascade that can be distinguished from the general fall. The gorge separates the Plain of Jaulan, Golan, or Gaulonitis, from the Mountains of Galilee; and no doubt much light that is now wanting, will be thrown upon the juxta-position of these features by the extension of the Survey to the east of Jordan. The western mountain supplies four streams to the northern shore of the Sea of Galilee, and it culminates at the height of 2,761 feet, on the east of Safed. It is indeed a prolongation of the Southern Range of Upper Galilee, which rises in the Plain of Acre and runs eastward, dividing Upper and Lower Galilee; while here at its eastern end the range separates the Huleh Plain, which is above sea level from the Sea of Galilee, which is 628 feet below the sea level. The range thus marks the commencement of the great depression that carries Jordan down to the Dead Sea, where the surface of the water or brine is 1,292 feet below the Mediterranean, the depth of the water or the soundings being about the same, making the total depression about 2,600 feet. It is along this extraordinary feature that the eastern base of the Western Highland of Palestine will now be examined.

THE WESTERN SHORE OF THE SEA OF GALILEE.

At the entrance of Jordan into the Sea of Galilee, the Plain of Batihah is spread out on the east at the foot of the Plateau of Jaulan, while on the west the hills descend with a rugged slope to the river and the lake. In this plain is the site of Bethsaida Julias. The path westward is at a little distance from the water. About two miles west of the river in an open situation at the foot of the slope, are the ruins now called Tell Hum, one of the sites attributed to Capernaum; Robinson regards it as Chorazin. Conder, following Neubauer, suggests Caphar Ahim, a town named with Chorazin in the Talmud.*

A wady enters the sea on the east of Tell Hum, and leads upwards to Kh. Kerazeh on the hill-side, about two miles and a half from the sea. This site is identified with Chorazin. About a mile and a half along the shore, westward of Tell Hum, is the small plain of et Tabghah or Tabighah, identified with Bethsaida of Galilee. It is a fishing village to this day. A cliff or rocky promontory projected into the sea, obstructs further progress along the shore, and divides et Tabghah from the Ghuweir or the Plain of Gennesaret.

The Plain of Gennesaret extends for three miles along the shore of the lake between the rocky heights and promontory which terminate it on the north, and the lofty cliffs of Wady el Hamam on the south. It recedes in a gradual curve from both ends, until it becomes a mile and a half in breadth. It is crossed by Wady 'Amûd, Wady er Rubudiyeh, and Wady el Hamâm. These embrace the drainage of the highlands between el Jish and Hattin. There is also a stream from 'Ain el Mudauwerah which rises in the plain. The ruins around Khan Minia at the northern end of the plain, are considered by high authorities to be the site of Capernaum. The present village of el Mejdel at the southern end, represents ancient Magdala. Mons. Guerin would place the ancient Chinnereth at Abu Shusheh.

South of Mejdel the highland advances to the sea shore, and a narrow track follows the rocky slope around a projection, which shuts out the view of the plain on the north. A mile and a half from Mejdel the southward track enters the rich valley of Wady Abu el 'Amîs, with the fountains of Ayun el Fuliyeh near the shore. The wady comes down from Merj Hattin, and from the rear of the lofty cliffs that dominate the Wady el Hamâm or Dovedale, together with the plain beyond. On the summit of the cliffs, and overlooking Wady Abu el 'Amis is Kh. Irbid or Arbed, the

* Neubauer, "Geog. du Talmud," 220, 221. Conder's "Tent Work," ii, 183.

Arbela of Josephus (Ant. xii, 11, 1), and the Maccabees (1 Macc. ix, 1, 2) and the Beth-Arbel of Hosea, x, 14. Near Irbid is the Kulat (Castle) Ibn Mân, the castellated or fortified caverns in the face of the lofty precipices of the Wady el 'Hamâm. They were besieged and taken by Herod, whose soldiers were let down in boxes suspended by chains from the top of the cliff, as it was found impossible to scale the cliffs from the bottom. They were afterwards fortified by Josephus (Josephus, "Wars," I, xvi, 4, "Life," 37). Dr. Tristram and his party got access to the caverns, after Herod's fashion. "Land of Israel," 448. Mons. Guerin reached them by means of steps cut in the face of the rock and communicating with corridors and galleries in successive stages. Guerin, "Galilée," i, 201–203.

On the southern side of Wady Abu el 'Amîs, facing Irbid, is the Hajâret en Nusâra, a clump of basaltic blocks, on the waterparting between Wady Abu el 'Amîs and Wady Fejjas, which drains the Plain of Alma. This is the reputed site of the miraculous feeding of the five thousand; although according to the circumstances of the narrative, it occurred on the eastern side of the lake. About two miles on the north-west along the same ridge are the Horns of Hattin, twin peaks, which the Latin Church has chosen for the place of "the Sermon on the Mount." It is more certain that the decisive Battle of Hattin took place in 1187 on the plain between the Kurn or Horns of Hattin and the village of Lubieh.

The road from the springs of el Fuliyeh also called 'Ain el Barideh, skirts the shore at the foot of the hills, and after about a mile and a half enters the northern gate of Tiberias. A path runs parallel with the road, on higher ground along the hill-side, and enters Tiberias by the western gate. The town is surrounded on the land side by a wall, strengthened by many towers and a castle, but breached in various parts by the earthquake of 1837. It lies along the shore for half a mile, and has a width of a quarter of a mile. The ancient city was of much greater extent. The present town is situated at the northern end of an uneven plain, extending along the seaside for about a mile, and running inland for a little more than half a mile at its widest part. The famous baths of ancient Emmaus or Hammath are a mile and a half south of the western gate of the town. In this direction the mountain rises close to the water in lofty cliffs, which attain to an altitude of 1,650 feet above the lake. The highland continues to hug the shore, and bold cliffs descend to the water, the road passing over them until the end of the lake is approached, and also the re-commencement of the Jordan. As the shore deflects slightly to the east, the mountain turns in a similar degree to the west, throwing off, however, a small spur to the mouth of the river. The spur becomes a well-defined mound along the shore, almost cut off from the land by a backwater of the river. On this mound are the ruins of Kerak, the Tarichæa of Josephus.*

The Jordan issues from the Sea of Galilee on the western side of a plain, about five miles in width, and extending southward, probably without diminution, for about 14 miles, when it expands into the great terraced recess of Beisân, for 11 miles further south, or as far as its junction with the Wady el Maleh. Below this point, the Jordan enters a gorge, and continues in it for about 12 miles, that is, as far as its junction with Wady el Bukeia ; here it begins to expand very gradually into the great plain, often diversified with low hills, which ultimately becomes the Plain of Jericho, and terminates on the south at the Dead Sea. Along the western margin of the Dead Sea, and at the base of the lofty cliffs which overhang it, there is generally a narrow strand sometimes expanding to a mile in width and seven to two miles and sometimes blocked altogether by cliffs advancing into the sea.

The following description specifies more particularly the variations which the Valley of the Jordan undergoes, on the south of the Sea of Galilee. It will be confined to the western side of the river and of the Dead Sea, that is, to the limit of the

* Macgregor's " Rob Roy," 408, 413.   Guerin, " Galilée," i, 275.

Survey, because the inaccuracies and defects in the best of former maps brought to light by the new work, gives to the information in existence concerning the eastern side, an insufficient and defective character.

The Ghôr, from Kerak to Jisr Mujâmia.
That portion of the broad plain beginning on the south shore of the Sea of Galilee, which is found on the western side of the rivers, varies for about three miles, from a quarter to three quarters of a-mile in width, in accordance with the approach or recession of the stream to and from the foot of the western heights.

About a mile from the sea, the western plain is crossed by the permanent stream of Wady Fejjas, which emerges from a rocky gorge, with cliffs at el Kulah on the south, that rise to 1,840 feet above the depressed plain, or 1,179 feet above sea level. The cliffs on the north are probably lower. An ancient aqueduct passing from the gorge along the hill side, conveyed the water of the Fejjas to Tiberias. The gorge is only a mile or two in length, and forms the avenue to the great plain, or Sahel el Ahma, which stretches westward from the heights that overhang Tiberias and the Sea of Galilee, as far as Hattin and Lubieh. It is very fertile, but devoid of the picturesque, and monotonous.

At the junction of the Fejjas with the Jordan, but on the opposite bank, is the village of Umm Jûnieh, with the remains of a bridge on the north. Further south is the village of el 'Abeidiyeh, on the western side of the river. From this place there is the main road to the north and south, and another which passes through the gorge of Wady Fejjas to meet, near Kefr Sabt, the road between Tiberias and Acre.

About a mile south of el 'Abeidiyeh, the plain is closed by the advance of a bluff towards the river; but the passage of the Jordan to the south-east, and the open nature of the ground about the small Wady Umm Walhân, helps to restore and enlarge the plain up to the confluence of Nahr Yarmûk, from the east, the River Hieromax of antiquity. This considerable affluent has had the effect of forcing the Jordan back towards the west, but it quickly turns east for half a-mile, and then takes up its former direction southward, for a short distance, and then returns south-westward to meet another advance of the hills, which close upon the river near the important passage of the Jordan, by the Bridge or Jisr Mujâmia.

The Ghôr from Jisr Mujâmia to Nahr Jâlûd.

On the west of the bridge, the hills fall back, and give this end of the plain the width of a mile, which expands at two miles farther south, to a width of two miles, chiefly owing to a remarkable bend of the Jordan, due east for more than half a-mile.

This easting of the river is maintained, and upon the whole increased, in course of the next three miles, when the river makes another sharp bend, and runs for half a-mile somewhat north of west. This bend is the beginning of a reaction of the Jordan towards the west, amounting to a mile and a half of westing, in two miles and a half of southing, the extreme point westward and the commencement of a new return eastward, occurring at the junction of the Nahr Jâlûd, which rises on the edge of the Plain of Esdraelon, and passes the ruins of Beisân.

Up to this point the western plain has a length of eight miles, from the beginning of this section near Jisr Mujâmia; and its width is, for the most part, about two miles. Besides the Nahr Jâlûd at the southern end, two other important permanent streams cross this section of the plain, namely—(1) the Wady Bireh, which falls into the Jordan at a mile and three quarters south of Jisr Mujâmia, and (2) the Wady el 'Esh-sheh, which joins the main stream four miles lower down.

Wady el Bireh enters the plain from a magnificent gorge, not remarkable for cliffs, but for its depth, and for the grand fortress which crowns the commanding summit on the south of the gorge, rising to a height of 1,850 feet above the plain, at the junction of the streams. The fortress, now in ruins and only occupied by a few poor wretches, is named Kaukab el Hawa, and was the crusader's Castle of Belvoir or Belvedere. The wady descends from Mount Tabor, about 10 miles from the mouth of the gorge, and passes through the north-eastern extremity of the Plain of Esdraelon, which is at the head of the gorge, only six miles from the Ghôr or Plain of the Jordan.

Wady el 'Esh-sheh emanates from the hills on the north of Beisân, and of the Valley of Jezreel, and rises on the east of Jebel Duhy. About a mile south of its confluence with the Jordan, is the ford of 'Abârah, which Lieutenant Conder identifies with Bethabara of John i, 28.*

In this part of the Ghôr, the lowest level of the Jordan Valley, in which the channel of the river meanders, has not yet become distinctly separated from the higher plain; but indications of such a feature begin to be displayed. Immediately on the south of the Nahr Jâlûd, this feature becomes distinct and prominent. There, the river's winding bed is sunk in a deep flat of varying breadth, called the Zor, which is enclosed between steep and sometimes perpendicular banks, as much as 150 feet in height, if not more, on the western side, to which the limits of the Survey confine these remarks; the river occasionally washes the foot of the high bank, and elsewhere it is as much as a mile away, and repeatedly half a mile, the low flat of the Zor taking its place. This precipitous bank is the limit of the plain, that extends from the top of it to the foot of the mountains. It is this upper plain that is called the Ghôr, in distinction from the bottom flat or Zor. Commonly the whole valley, from the Lake of Tiberias to its southern end, bears the Arabic name of el Ghôr, or the depression. The distinction between the Ghôr (pronounced Rôr) and the Zor, was explained by Lieutenant (now Colonel) Charles Warren, in "Notes on the Jordan Valley," dated 21st October, 1868. It is also attested by former travellers.

* "Tent Work," ii, 64.

The Plain of Beisân.

The plain of Beisân may be said to extend from the Nahr Jalûd to the Wady Maleh, a distance of 11 miles. The Valley of the Jordan here undergoes a great widening, which commences at the outfall of Nahr Jâlûd, into the Jordan, and from thence stretches westward along the foot of the hills on the north of Beisân, and up the Valley of Jezreel, to the waterparting between it and the Plain of Esdraelon, or Megiddo, or Merj ibn 'Amîr, about 15 miles from the Jordan.

On the south side, the Valley of Jezreel is bounded by the mountains of Gilboa (Jebel Fukû'a), until they come to about three miles west of Beisân, when the mountain range turns to the south, and forms the western limit of the Plain of Beisân. For about seven miles south of Nahr Jâlûd, the plain on the west of the Jordan is six or seven miles wide; beyond which low hills advance suddenly, and reduce this side of the plain to a width of about three miles at its southern end. On the eastern side of Jordan, the foot of the mountains of Gilead runs straight from north to south, and is two miles distant from the river. Among these highlands are Jabesh Gilead and the remains of Pella.

Another abrupt rise takes place in the plain, like that which confines the Zor. It occurs along the length of the broad part, from Nahr Jâlûd and Beisân on the north, towards the advancing hills on the south, where they are capped by Khurbet Kâ'aûn. It lies midway between the Zor and the foot of the western mountains, and is parallel to them. The height of this remarkable bank, is more than double of that which confines the Zor, and amounts to 400 feet or more. Beisân is on its edge, and so is the Roman road which connects Beisân with Nablus. It will be seen that this striking feature, although it stands far in front of Mount Gilboa, is nearly in a line with the hills at its northern and southern extremities, and thus appears to be the proper western limit of the Ghor, separating it from the long and gradual slope which the Valley of Jezreel makes in ascending to the Plain of Megiddo.

The following altitudes define the variations in height which take place in the Plain of Beisân. On the north side of the plain, at the confluence of the Nahr el Jalûd with the Jordan, the Zor may be reckoned from collateral observations, in the absence of any on the spot, to be 930 feet below sea level. At the junction of Wady Maleh with the Jordan on the south of the plain, the Zor is 1,060 feet below sea level. The altitudes that have been observed in the Ghôr, or the first step above the Zor, range from 850 to 685 feet below sea level. The altitudes on the edge of the second step upwards, rise from 426 to 322 feet below sea level. At 'Ain Tub'aûn, in the Valley of Jezreel, nine miles west of the edge of the second slope, the altitude is 120 feet *below* sea level; and at the head of the valley, on the edge of the plain of Megiddo, about four miles from 'Ain Tub'aûn, the altitude may be reckoned at 240 feet *above* the sea. The total fall from the eastern edge of the Plain of Megiddo to the Jordan, at the junction of Nahr el Jalud, is 1,170 feet; the distance being 15 miles. The fall is therefore at the rate of 78 feet per mile, measured in a direct line.

The hydrography of the Plain of Beisân, is obscured by the abundance of irrigation channels, spread nearly all over the upper and lower terraces. These unhappily now serve to denote present waste, and also the high development of productive industry in these regions in former times, together with the inducements thereto, which the soil and climate offer, in combination with a well directed artificial irrigation, which the cultivators themselves appear to understand, much better than theoretical engineers, whose tendencies are too one-sided, too much bent upon the works themselves, and too little identified with the varying incidents that must be met by the cultivator himself from day to day. Such appears to be the explanation of failures that have too often followed great irrigation works by professional Europeans in India, and the reflection suggested by such examples as this great irrigated plain, is, that native talent and administration under the oversight of supreme European governments, is the proper combination to secure, in the first place, protection for life' and property; and in the second place, the highest results for local industry. Looking at the grave necessity for good administration in these long-oppressed regions, and the prospect of an early application of European instrumentality for their restoration to wealth and prosperity, that shall benefit, not these regions alone, but all the hives of manufacturing industry elsewhere; it does not seem out of place to seize upon such an occasion as this to strike a note that may, by the providence of the Almighty, reach home in the right direction.

The water derived from the Nahr Jâlûd, and from numerous fountains and brooks that descend from the hills, feeds the irrigation still partially in use, as well as the marshy tracts that denote the extent of the waste enforced by Arab spoliation and Turkish oppression. From these sources several streams pass from the upper to the lower terrace, and from the latter to the Jordan in the Zor. As many as six permanent streams thus find their way to the Jordan, between Nahr Jâlûd and Wady Maleh. The streams descending from Jebel Fukû'a, have little room for development on the short slopes; but where the hills encroach on the southern part of the plain, the Wady Shubash and the Wady el Khashneh, drain more important valleys. The Shubash indeed rises on the western side of Jebel Fukû'a, and cuts through the range. The Khashneh rises in Ras Ibsik (altitude 2,404 feet), and skirts the Roman road to Beisân and Nablus.

All over the upper and lower terraces of the great plain, numerous tells or mounds are distributed; they still bear distinctive names, and are the sites of former habitations, scenes of domestic happiness, and abundant wealth, that may be restored almost as rapidly as they were obliterated, when once the civilization and power of the West becomes conscious of the connection between Oriental prosperity and that of its own manufacturing populations.

These tells probably mark the substantial and lordly centres of villages, the latter more or less extensive, and readily levelled with the ground. They denote the populous character of the region, when a strong government restrained the plundering Ishmaelite, and protected instead of robbed the people. The tells are more indicative of a large population than the remains of such a "splendid" and "noble" city as Beisân, when it was either Jewish ¡Beth Shan, or heathen Scythopolis, with its dominating citadel, temples, hippodrome, theatre, baths, monuments, and bridge.*

At the southern end of the plain, are the ruins near 'Ain es Sakût, which Mons. Guerin identifies with Succoth of 1 Kings vii, 46, and 2 Chron. iv, 17. But he carefully distinguishes this site from the Succoth of Jacob in Genesis xxxiii, 17, and of Gideon, Judges viii, which is identified with the town of that name, given to Gad, on the east of Jordan, Josh. xiii, 27.†

This eastern Succoth is called Taréla, or Tarala, in the Jerusalem Talmud.‡ Jacob sojourned there on his way from Haran in Syria to Shalem of Shechem, passing the River Jabbok. Gen. xxii, xxiii. Where the Jabbok emerges from Mount Gilead on to the plain of the Ghor, here broad and ample, the mound of Deir Ula, Darala or Tarala, commands the plain. Its discovery is ascribed by Lieutenant Conder to the Rev. Selah Merrill.§ But the site was mapped by Lieutenant Warren, and catalogued in his report of October 19th, 1868, which was published in the papers of the Fund at the time. It was inserted in the map of the Holy Land, in Dr. Smith's Ancient Atlas, from Lieutenant Warren's sketch.

At the foot of Mount Gilboa, and three miles south-west of Beisân, is the ruin called Khurbet el Mujedda, which has been adopted by Lieutenant Conder, as the site of Megiddo, in preference to the position on the Mukutt'a, near Lejjûn.|| Besides the objection to this proposition, derived from the

* Robinson, iii, 329-332. Guerin, "Samarie," i, 285-298.
† Guerin, "Samarie," i, 269.
‡ Neubauer, "Geographie du Talmud," 248.
§ Conder's "Handbook to the Bible," 253.
|| "Tent Work," i, 128, ii, 68.

separation of the site from the River Kishon, and the town of T'anak, there is the account of the flight from Jezreel, of Ahaziah, King of Judah (2 Kings ix, 27), which it seems impossible to reconcile with such a position for Megiddo. Ahaziah fled by way of the garden house (Beth-hag-Gan or Bethgan), which Dr. Grove, Mons. Guerin, Dr. Tristram, and Dr. Stanley, identify with Engannim, the modern Jenin. Dr. Stanley observes that "the garden-like character of the spot is still preserved."* The king was smitten at Maaleh Gur, "the going up to Gur," near Ibleam, and he fled to Megiddo, where he died. It is difficult to conceive how this narrative can be reconciled with a Megiddo in the Plain of Bethshean; for it must be remembered that Jehu, the pursuer of Ahaziah, made his furious advance upon Jezreel, through that plain; and it seems highly improbable that the flying monarch would have rushed towards his opponent, instead of from him. This difficulty alone seems insurmountable. But there is also Engannim or Bethgan, and the other points in the story, to be sought for somewhere in the Valley of Jezreel; but nothing except Mujedda, suitable for the purpose, has yet been discovered in that direction.

The Samaritan Gorge of the Jordan.

The broad Plain of Beisân is succeeded by a gorge or narrow valley, extending about 11 miles between Wady Maleh, and Wady Abu Sidreh, the latter being the lower course of Wady el Bukei'a. This is the narrowest part of the Lower Jordan, and according to Lieutenant Warren it contracts to not more than half a mile in width. "For six miles the plain is nearly lost." Guerin repeatedly calls it "très étroite." According to Conder it is two to three miles wide.

The summits of the hills which enclose the Ghôr on the west, rise above the Jordan, to heights between 1,100 to 1,840 feet, their heights above the sea being about 1,000 feet less.

This range is the eastern boundary of the Wady Maleh

* "Sinai and Pal.," 349, note.

basin, which drains the hills that form the western side of the gorge. But the principal affluents of the Wady Maleh are derived from the range that bounds the basin on the south-west, and constitutes its main slope or watershed. One of these affluents runs for seven miles northward, along the eastern side of the basin to join the Jordan, which descends on the eastern side of the same range, in a direction exactly opposite.

This south-western range seems to be the chief factor in the group of hills that forms the western side of the Samaritan Gorge. It was observed, heretofore, that the range forming the axis of the hills on the western side of the Galilean Gorge of the Jordan, is in continuity with the mountain range that divides Upper and Lower Galilee, and extends to the Plain of Acre. In like manner the southwestern range of the Maleh basin, forming the main element of the hills on the west of the Samaritan Gorge, exhibits a remarkable alignment with Mount Carmel, and its continuation along the south-west of the Plain of Esdraelon. This range is also the first of a series of ranges parallel to it, which will be noticed hereafter.

The foot of the hills on the west of the gorge, often descends to the Ghôr in precipitous rocks, and advances occasionally to the Zor, the track being carried sometimes along the very foot of the hills or over the slope of the projected spur, or even down into indents of the Zor. For the Zor itself very much indents the Ghôr, penetrating it along the course of the wadys, which descend in ravines across the Ghôr from the hill-side. The Zor also varies in width on the western side (of which alone the survey takes cognizance), independently of the indents before-mentioned. It exceeds a mile at Wady Marma Fiad on the north, and also at Wady Abu Sidreh on the south, where it reaches up to the hills; and it is quite half a-mile broad for nearly two miles, on the south of Wady Umm ed Deraj ez Zakkûm. Ten fords across the Jordan, are noted and named in the new Survey within this narrow tract.

THE SAMARITAN GORGE. 161

In the reconnaissance map and report of this gorge made by Lieutenant Warren, R.E., in October, 1868, various names of wadys occur, which would probably supply those which are omitted in the new map. Mons. Guerin also contributes some names and objects here, as elsewhere, that do not appear on the map.* The extension of the Survey to the east of the Jordan, will afford an opportunity of reconciling or explaining these discrepancies upon further inquiry on the spot.

A COMPARATIVE TABLE OF NAMES FROM WARREN, GUERIN, AND CONDER, BETWEEN W. MALEH AND W. ABU SIDREH.

FROM WARREN'S SKETCH, 1868.	FROM GUERIN.	FROM CONDER'S SURVEY.
W. Malih.	O. el Maleh.	Wady el Mûleh (esh Sherar).
W. Um Karuby.	O. Marmy Faiadh.	—— Marma Fiad.
W. el em Dahideh.	O. Rhazal.	Thogret el Kabur.
W. Shiyeh.	O. es-Seder.	Wady Fass el Jemel (Habs Katurj).
W. Swaida.	Sath er Rhoula (cave).	
W. Saujeh (S'aidiyeh).	Kh. el Bridje.	—— Shaib.
W. Abu Serad (Jerad).	O. es Seka'ah.	—— Umm ed Deraj ez Zakkum.
W. Belgôu.	Haouch ez Zakkoum.	
W. Abu Haschish.	O. en Nekeb.	Un-named stream.
W. Ghor.	Tell es Saidieh.	Sadet et Taleb.
W. Abu Sidra.	O. Asberra.	Wady Jurat el Katufi.
	O. Abu Sehban.	Three ruined sites.
	O. Kefr Anjda.	Arak Abu el Hashish (rocky eminence above unnamed stream).
	O. ez Zarha.	
	Un autre Oued.	
	O. el E'urkan.	Wady Abu Loz.
	Siret el Maazeb.	Sidd el Belkawy. *See* Belgôd above, in Warren.
	Kh. es Sireh.	
	O. Abou Sedra.	Tombs, cisterns, and Kh. el Karur.
		Tell and Wady Abu Sidreh.

The Ghôr between the Gorge and the Dead Sea, in 2 parts. South of Wady Abu Sidreh, where the Samaritan Gorge is considered to terminate, the Ghôr begins to widen by successive accretions, and attains a breadth of five miles west of the Jordan, where the Wady el Humr crosses the plain, on the north of Khurbet Fusail, the ancient Phasaëlis. This width continues undiminished up to the Dead Sea; but in other respects the Ghôr exhibits a remarkable variety of features which makes it desirable to divide the plain on the south of the Samaritan Gorge, into two parts, the separation of which is found along the southern waterparting of Wady el 'Aûjah Basin. The northern part may be called the Plain of Phasaëlis, after the Herodian city, the ruins of which remain in its midst. The southern part is the famous Plain of Jericho.

* Guerin, "Samarie," i, 264–268.

1. *The Plain of Phasaëlis.*

The Wady Abu Sidreh was chosen to define the southern limit of the Samaritan Gorge, because it is a line that lies distinct upon the ground, and is equally clear upon the map. But the waterparting on the north of this wady should be regarded as the actual limit, although it may not be so easy to trace. The preference for the waterparting is due to natural features connected with the widening of this part of the Ghôr. These occur in the form of a series of parallel valleys, commencing on the south of the range that bounds the Maleh Basin on the south-west, and which has been shown to have a remarkable influence on that basin, and also on the Samaritan Gorge. These valleys are (1) the Wady el Bukei'a. (2) the Wady Fâr'ah, and (3) the Wady el Ifjim, Lakaska, or el Humr. They run in parallel courses through the hills into the Ghôr, and, with a singular exception, enlarge the Ghôr by the deep bays which run up from it into the hills, and form the mouths of these valleys.

The Wady Sidreh is the outlet of the first of the series, the Wady el Bukei'a, which runs at the base of the range that bounds the Maleh Basin on the south-west. The Wady el Bukei'a is, however, somewhat exceptional at both ends. At its head the Bukei'a is overlapped by the broad expansion of the sources of the next parallel valley which succeeds it on the south,—the noted Wady Fâr'ah, the great highway between Nablus and Trans-Jordan. At its outlet the Bukei'a takes a very curious and singular course; for instead of gradually expanding into the wide plain, and so contributing to the surface of the Ghôr, it contracts into a narrow, deep, and rocky chasm, which appears at first to be completely blocked at its lower end, while it really makes an abrupt and short bend nearly at right angles to its former south-westerly and straight course, and then winds about northward, north-east, and finally east, till it emerges from the hills to descend at once into the Zor, which here interrupts the Ghôr altogether.

So little is there any obvious connection between the

Wady el Bukei'a and the Ghôr, that they might be superficially considered to be quite distinct. But no one will deny the connection of Wady Fâr'ah with the Ghôr, in the fertile and well-watered tract of Kurawa, which is at once the mouth of Wady Fâr'ah, and a noted part of the Ghôr. Nor can the commencement of the series of parallel valleys which include Fâr'ah, fail to be seen in Wady el Bukei'a.

For five miles south of Wady Abu Sidreh, the Ghôr west of Jordan is hemmed in by the end of the range that divides Wady el Bukei'a from Wady Fâr'ah; and here the Ghôr nowhere exceeds two miles in width. The range terminates in a point at el Makhrûk, near ruins which are probably the site of Archelaus; and the Ghôr at once expands into the Kurawa at the mouth of Wady Fâr'ah, and attains a width of four miles. The Kurawa runs up north-west of Makhrûk for four miles, and there terminates in a rocky chasm, through which the Wady Fâr'ah descends from the broad valley above.

Next to the Valley of Jezreel, the Wady Fâr'ah forms the most open and important avenue between the Jordan and the West; and a full description of it has been given in pages 76 to 81.

The southern or south-western side of the Kurawa and Wady Fâr'ah, is the range which divides them from Wady Ifjim, which enters the plain as Wady Zakaska, and crosses it as Wady el Humr. The range terminates in a spur from the remarkable peak of Kurn Surtabeh (alt. 1,244 feet or 2,390 feet above the Jordan). At the southern base of this mountain, the Ghôr expands to a width of five miles, and runs up into a tapering recess for about three miles on the south-west of Kurn Surtabeh. The recess is a mile and a half wide at its mouth. It is bounded opposite Kurn Surtabeh, by lofty cliffs which form the northern termination of the long line which, like a wall, forms the base of the western mountains, and extends with little interruption, to the ascent of Akrabbim, at the south of the Dead Sea, a distance of about 80 miles.

Into the recess descends the third in the series of parallel valleys, the Wady Ifjim. The passage has a peculiarity rivalling that of the outlet of el Bukei'a, the first of these valleys. The head of the recess is divided into two by a narrow spur about a mile long, which projects southward from the mountain, and has a gorge on both sides, the western being a rocky chasm, one side of which is the beginning of the long line of cliffs already mentioned. The Wady Ifjim comes down from a beautiful plain about two miles north of the chasm, through a rocky cleft, as if it were going to descend through the chasm, or parallel gorge on the east of the spur. But instead of doing so, the wady deflects slightly to the westward, and comes out into the plain by a chasm from the west, which meets the northerly chasm at its outlet. A Roman road which passes from the Ghôr, ascends by the spur aforesaid, and skirts the eastern side of the wady into the Sahel or Plain of Ifjim, and so onward to the Plain of Salim and Nablus.

In another respect the Wady Ifjim is a counterpart of the Bukei'a. For as the head of the latter is overlapped by an expansion of Wady Fâr'ah towards the north, so is Wady Ifjim overlapped by another expansion towards the south, as follows:—From the Sahel or Plain of Ifjim, there is a continuous ascent by the Wady el Kerad to the elevated Plain of Salim, which runs in the same line, but drains in a contrary direction to the Wady Fâr'ah. *See* page 78. Another ascent from the Plain of Ifjim proceeds westward by Wady Zâmûr to Yanûn, and thence by a valley in the same line, but draining in the opposite direction, to the Plain of Mukhnah, which is a continuation of the Plain of Salim, but in another basin.

About a mile and a half south of the base of Kurn Surtabeh are the remains of the Herodian City of Phasaëlis, now Khurbet Fusail. The site is on a wady of the same name which joins Wady el Humr in the Zor. The wadys on the south belong to the basin of Wady el 'Aûjah, which drains the southern division of this part of the Ghôr, up to the commencement of the Plain of Jericho.

THE PLAIN OF PHASAËLIS. 165

The mountains maintain their uniform direction southwards for three or four miles below Fusail. Low hills then advance eastward into the Ghôr, up to a frontage corresponding with the eastern side of Kurn Surtabeh. Between five and six miles from Fusail, the mountains recede westward for about two miles from their former line, and become separated from the low hills by an enclosed plain, varying in width up to about two miles. The low hills run southward for about seven miles, and their termination is distinguished by a prominent summit called 'Osh el Ghurâb or the Raven's Nest. The Ghôr now recovers its full width, and becomes the Plain of Jericho. Towards the end of the hills, the mountains also gradually advance up to their former line with rocky precipices at their base; and leaving only a narrow entrance between the enclosed plain and the Plain of Jericho. In the entrance is the noted fountain of 'Ain Dûk and 'Ain en Nuei'ameh. The length of the enclosed plain is about four miles. Within this brief distance, the plain is crossed by Wady el 'Aûjah, Wady el Abeid, and Wady Umm Sirah, and these also intersect the hills on the east in proceeding to the Jordan by the Wady el 'Aûjah. It seems probable that the Vale or Plain (Emek) of Keziz, belonging to the tribe of Benjamin (Josh. xviii, 21), may be identified with this enclosed tract.

To the Ghôr itself, the name of "the Plain of Phasaëlis" has been applied. Among the features which particularly characterise this part, may be noted the remarkable manner in which the Wady Fâr'ah as Wady el Jozeleh, and the Wady el Mellâhah, a branch of Wady el 'Aûjah, both advance across the Ghôr from the west, to within a mile and less of the Jordan, and then turn southwards, pursuing a parallel course to the main stream for six or seven miles, before the confluence of each is effected. Observations are wanting to denote the relative altitude of the tributary beds to the bed of the main stream, or to the surface of the Ghôr; but it is evident that these parallel passages of the tributaries produce a considerable disturbance of the surface of the Ghôr, degrading or wearing it away to a much greater extent than

M

is found along the ordinary embankment of the Zor. Perhaps the unusual course of these river beds may have been originally due to artificial channels made for the purpose of irrigation. The narrow tongues of land between the tributaries and the Jordan are called in Arabic, Monkattat or Mankatta'at, which according to Professor Palmer appropriately means "strips."

On the south of the outlet of Wady el 'Aûjah, the Ghôr is watered by the branches of Wady Mesâ'adet 'Aîsa, which rises in the low hills crowned by 'Osh el Ghurâb.

The Arabic name of this wady means the "Ascension of Jesus," so called with reference to the conical hill of 'Osh el Ghurâb (Raven's Nest), the traditional site among the Arabs for the mountain of the Temptation.* Christian tradition originating in the time of the Crusaders, places the same event at Jebel Kuruntul or Mount Quarantana. On the north of the wady are two groups of ruins with extensive sandstone quarries, called es Sumrah, which are commonly identified with Zemaraim, a Benjamite City. Josh. xviii, 22.

The next watercourse is Wady en Nuei'ameh, which crosses the Ghôr from 'Ain Dûk, and divides the Plain of Phasaëlis from the Plain of Jericho.

2. *The Plain of Jericho.*

The Ghôr acquires a comparatively unbroken surface in the Plain of Jericho. Its commencement on the north is at Wady en Nuei'ameh, and its termination on the south is defined by the embankment of the Zor, which also comes to an end on the north of the Dead Sea. At Kusr el Yehûd or Jews' Castle, about two miles below the confluence of the Nuei'ameh, and nearly eight miles from the Dead Sea, the embankment begins to slope away from the Jordan in a south-westerly direction towards the foot of the western mountains, narrowing as it goes, till it ends in a point at Khurbet Kumran, which is ten miles south of the north-west angle of the plain.

* Conder's "Tent Work," ii, 13.

At its mouth the Jordan prolongs its western bank into a point advanced into the sea for half a mile farther south than its eastern bank. From this point the coast recedes north-westward into a bay, and bending round to the south-west contracts the Zor, here forming the sea-shore, to a width of a mile, and finally brings it to a point at Ras Feshkah, about two miles south of the end of the Ghôr at Khurbet Kumran.

The lofty and precipitous rocky base of the western mountains enters the plain from the north-west, and preserves that direction as far as Jebel Kuruntul or Mons Quarantana, the reputed scene of Our Lord's Forty Days' Fast. This mountain is about a mile south of the north-west corner of the plain, and from thence the line of cliffs takes a more southerly course for a mile and a half, when Wady Kelt comes down through them. The modern village of Jericho (Erîha) lies two miles east of the gorge of the Kelt, and the ruins of the successive sites of ancient Jericho extend towards the gorge, and also towards the north-western extremity of the plain, especially about 'Ain es Sultan, a great fountain not quite a mile east of Jebel Kuruntul. Wady Kelt is identified with Elijah's Brook Cherith ; and 'Ain es Sultan is deemed to be the fountain which Elisha healed. 2 Kings ii.

On the south of Wady Kelt, the cliffs subside, and while the foot of the hills, in running southward, advances slightly towards the east, the summits retire in a semicircle towards the west, giving an easier slope to the spurs as far as the Pass of Kueiserah. At the pass, the cliffs reappear, and run on with a slight ogival curve to Ras Feshkah, where they enter the sea, and divert the passage along the shore to a path over the cliff.

Although the Ghôr appears to be perfectly level, the instrumental observations of the Survey have proved that it has a slope of some 500 feet between the foot of the cliffs, on the west of Jericho, and the edge of the descent to the Zor, near the ford of el Henu, a distance of six miles. At Kh.

Kakûn near Wady Kelt, the alt. is 585 feet. At Kusr Hajlah, the alt. is 1,066 feet. The Kusr Hajlah is a mile and a half from the descent to the Zor. The channel of the river at el Henu, a mile from the descent, is 1,254 feet, or about 150 feet below the edge of the Ghôr.

Except at its north-west angle, the plain of Jericho is now for the most part in a desert state; and yet it is naturally one of the most fertile and productive in the world, well watered by the rains of heaven, and having every facility for abundant irrigation, a soil that is described as "fertility itself," and a climate so favourable that Dr. Robinson found the barley fully gathered and threshed by the 22nd of April, and the wheat, of fine quality, was nearly harvested on the 13th of May.* Dr. Thomson found the barley harvest over on the 1st of April, and he notes that it comes on about the middle of March.†

Besides Wady Nuei'ameh on its northern boundary, the Plain of Jericho is crossed by Wady el Kelt, with an affluent from 'Ain es Sultan, which unites with the Kelt on its left bank, about half a mile from Eriha, the present village of Jericho. Half a mile lower, another branch called Khaur Abu Dhâhy, rises at the foot of the hills near Wady Kelt, and joins the Kelt on its right bank. Opposite the junction on the left bank of the Kelt, are the remains, called Jiljulieh, which are regarded as the site of Gilgal. At the foot of the descent into the Zor, the Kelt also receives a stream from 'Ain Hajlah, a fountain which rises about a mile and a half from the edge of the Ghôr, and is identified with the site of Bethhogla. About half a mile south-west of the spring is Kusr Hajlah, being the remains of a monastery, on the verge of the more ancient site. On the edge of the Ghôr, less than a mile north of the Kelt, is Kusr el Yehûd or Jew's Castle, the ruin of a great monastery dedicated to St. John the Baptist, that dominated one of the Pilgrims' bathing places on

---

* *See* Robinson's "Bib. Res.," 1, 534–568.
† Thomson's "Land and Book," 619.

the Jordan. Its crypt and other substructures exist, and were found by Dr. Tristram tenanted by flocks of rock doves.*

The Wady Kelt crosses the Zor for rather more than half a mile to its junction with the Jordan. Near the junction on the south, is the Hajlah Ford, the Pilgrims' bathing place of the present day. According to Lieutenant Conder, both the Latin and Greek Churches regard it now as the scene of Our Lord's Baptism.† Dr. Robinson, and also Dr. Thomson, make this ford the Greek site; and place the Latin site near Kusr el Yehûd.‡ This is reversed in Baedeker's Guide. In the sixteenth century, the pilgrims resorted to the Ford of El Ghoraniyeh, where Wady en Nuei'ameh joins the Jordan on the right bank, and Wady Nimrin on the left. This position is preferred as the site of Our Lord's Baptism, by M. Guerin,§ and by Dr. Tristram.‖

The Khaur el Thumrâr, rises in many branches on the western hills on the eastern side of Wady Talat ed Dumm. It crosses the Plain of Jericho on the south of the Hajlah road, near the Kusr Hajlah. On descending to the Zor, it bends to the south, and follows a course parallel to the Jordan into the Dead Sea. The Wady Makarfet Kattûm, and the Wady Joreif Ghuzal come down from the same range and cross the plain in courses parallel to the foregoing. They are succeeded by Wady el Kaneiterah, which rises on the eastern side of the Mount of Olives, and meets the head waters of Wady Kelt, near Shafat, on the north of Jerusalem. The Kelt again meets the head waters of Wady Nuei'ameh at Bethel, and the Nuei'ameh Basin reaches up to Tell 'Asûr. Thus the wadys which cross the Plain of Jericho, are found to form an important part of the watershed, including several roads to the highland between Jerusalem and the north of Bethel, as already explained in treating on the basins.

Further south, the Wady Kumran crosses the plain, where the Ghôr comes to a point at the ruins of Kumran,

---

* " Land of Israel," 225. † " Tent Work," ii, 17.
‡ Robinson, " Bib. Res.," i, 536. Thomson, " Land and Book," 615.
§ "Samaric," i, 114, 115. ‖ " Bible Places," 333, 334.

between the mountains and the Dead Sea, near 'Ain Feshkah. The wady descends from the elevated Plain of el Bukei'a, which contains other names like Kumran, which are severally and together very suggestive of a connection with Gomorrah.

THE WESTERN SHORE OF THE DEAD SEA.

Ras Feshkah to Ras Mersid.

Dr. Tristram's description of his bold examination of the coast between Ras Feshkah and Ras Mersid is hardly concurrent with the new Survey.* At Ras Feshkah, the plain is brought to an absolute termination by the descent of the headland into deep waters, rendering a passage impracticable even to an adventurous cragsman like Dr. Tristram, who had to scramble up and down the rocks and gullies away from the water line to reach the south side.† Lieutenant Conder pays high compliments to Dr. Tristram's map of the Dead Sea ; but there are discrepancies between the two maps which claim some explanation. In the older map, there is first, the bold bank projected into the sea on the south of Ras Feshkah, and next is the advance of the cliff to the shore on the north of Wady Derejeh, neither of which are supported by the New Survey. If these may be explained by a difference in the level of the waters, or otherwise, it would be well to do so.

From Ras Feshkah to Ras Mersid, a distance of 15 miles, the shore of the Dead Sea is restored, and gradually becomes about half a mile in width, at the foot of overhanging cliffs rising to a height of 2,000 feet. Eight miles south of Ras Feshkah, the strand projects into the sea, and becomes a mile wide, apparently from the detritus brought down from the mountains by Wady ed Derajeh (meaning steps), and Wady Husâsah (meaning gravel). Between the two great headlands, the cliffs form a slight curve which recedes for a mile and a half at Wady ed Derajeh, from a chord line stretched

* "Land of Israel," 276, 277.   † *Ibid.*, 254.

between the heads. The projection of the shore at this part, advances to the same chord line.

South of Wady Husâsah, the shore becomes narrow, and on the north of Ras Mersid, after passing Wady esh Shukf, a sulphur spring was discovered by Dr. Tristram on the shore at the foot of Ras esh Shukf (alt. of the Ras 1,227 feet, or 2,520 feet above the Dead Sea). The headland of Ras Mersid is only to be rounded with difficulty, and there is no track. Still it does not appear to be obstructive to all passage like Ras Feshkah, and it is presumed that deep water does not wash its base. A mile beyond Ras Mersid, another headland occurs on the north of Wady Sideir, and is crossed by the Nukb or Pass of Sideir, leading to the Plain of 'Ain Jidy (Engedi). Lieutenant Conder visited the sulphur spring from 'Ain Jidy. He appears to have got as far as Nukb Sideir on horseback, and then he had to dismount, "scrambling over cliffs or walking in the water round promontories," to reach the place.*

The principal wadys which cross the shore, and enter the Dead Sea between the headlands of Feshkah and Mersid, are Wady en Nar and Wady Derajeh. The Wady en Nar (Kidron) rises at Jerusalem, passes the monastery of Mar Saba in a profound ravine, and reaches the Dead Sea on the south of Ras Feshkah. The Wady Derajeh rises at Bethlehem, passes Jebel Fureidis, the site of the fortress of Herodium, and empties itself into the Dead Sea at the widest part of the shore, midway between the headlands.

The Plain of Engedi is about half a mile broad, and a mile in length. It has the cliffs of Wady Sideir on the north, and those of Wady el 'Areijeh on the south, while on the west rises terraced slopes on the top of which, 600 feet above the sea, is the plateau where the famous spring rises under a great boulder, and then falls down over the rocks to the plain below. Six hundred feet still higher, is another plateau in the form of a pentagon, on the summit of vast cliffs standing out with a salient angle to the south-east, like a bastion at the end of

* "Tent Work," ii, 137.

the mountain. This is the cliff of Zor (2 Chron. xx, 16). Plundering bands from the east, still use the pass on their incursions to the western highland. The plateau is surrounded by precipices on all sides, except at its throat on the northwest, where two roads unite, coming down the mountain side from the north and from the west, or from Bethlehem and Hebron. From this upper plateau, "the path descends to the seashore by ziz-zags, often at the steepest angle practicable for horses, and is carried partly along ledges or shelves on the perpendicular face of the cliff, and then down the almost equally steep *débris.*" Thus the lower plateau is reached. The further descent to the plain follows the course of the cascade derived from the spring. The water is almost hidden among the trees and shrubs and canes that flourish on its banks. The whole of this lower slope was once built up for terraced cultivation, and at its foot are the remains of ancient Engedi one among the oldest of cities.

The Wady Areijeh which bounds Engedi on the south is one of the principal drains of these mountains, and rises along the Hebron road, about Hulhul, and northward as far as Breikut and Tekua, which are the existing names of ruined sites corresponding to the Berachah and Tekoa of Jehoshaphat's deliverance. 2 Chron. xx, 20, 26. The Wady Sideir only comes from the heights above Ras Shukf, but the extreme beauty of the fairy scene in the lower part of the gorge is too attractive to be passed without notice. Dr. Tristram is enthusiastic in its praise.*

Shore of the Dead Sea—Ras Mersid to Sebbeh.

South of the Plain of Engedi, the shore is narrow, and follows the direction of the cliffs for two miles, with interruptions from gullies and boulders. At Wady el Kuberah, the line of cliffs inclines a little to the west of south, as far as Sebbeh, and the coast line becomes a mile distant from the great cliffs. The survey displays the projection of a considerable terrace extending between the foot of the cliffs and the sea,

* " Land of Israel," 289, 290, 297.

but Dr. Tristram mentions "four great rows of eroded terraces one above the other, and heaps of *débris* forming a slope at the foot of each."* The coast line recedes and advances in easy curves, from Wady Kuberah to Sebbeh, where the Survey ends ; and the plain gradually increases in width, until on the north of Sebbeh, it is two miles from the cliffs to the sea.

The isolated rock of Sebbeh rises midway between the line of cliffs on the west, and the sea on the east, and in the latter direction the summit of the rock is elevated above the plain for 1,500 feet. The altitude above the Dead Sea, according to the survey, is 1,702 feet, or 500 feet less than Dr. Tristram's barometrical measurement. The summit is a plateau 2,070 feet long by 1,050 feet broad; it bears the remains of the ancient Jewish fortress of Masada, built by Jonathan Maccabæus, completed by Herod the Great, and finally defended as the last refuge of the Jews under Eleazar.† Captain (now Lieutenant Colonel) Warren ascended by the remains of a zigzag track on the eastern face of the cliff, now almost impassable, which must be the remains of the "Serpent" path mentioned by Josephus. The easier ascent is by the western side, from which a ledge of rock is thrown off, and joins the western cliffs. A ravine runs along the southern scarp of this ledge, which slopes towards the north. Another ravine runs northward, dividing the slope of the ledge in that direction from the foot of Sebbeh. On the ledge, a great mound was piled up by the Roman besiegers against the western side of the fortress, for the purpose of planting their battering machines against the walls. This mound still affords access to the western entrance. To prevent assistance or the escape of the desperadoes in the garrison, the Romans built a fortified wall entirely around the base of the hill, and this, with the Roman camps outside, remains, and is traced on the survey.

Of the wadys which enter the Dead Sea in this part, the

* "Land of Israel," 302.
† Josephus, "Wars," vii, 8, 9.

Kuberah and the Seiyal rise on the waterparting that divides the tributaries of the Dead Sea from those of the Mediterranean, namely, Wady el Khulil, afterwards Wady Ghuzzeh. Along the ridge which gives rise to the sources of the Kuberah, are the noted sites of Beni Naim on the east of Hebron, Ziph, Carmel, and Maon. The basin of Wady Seiyal succeeds, and comes to a termination with the Survey at the biblical site of Arad.

In these pages, little more than a bare topographical description, explanatory of the Survey, is given of the Plains of Palestine. Their precise and accurate geography affords, however, scarcely any cue to the interest which belongs to them. That has already elicited the eloquence and learning of Dean Stanley, in his comparison of them with those of other countries,* and in his elaborate commentaries on the Maritime Plain, the Plain and Terraces of Jordan, and the Plains of Esdraelon, Hattin, and Gennesareth. These will be read with fresh zest in the light thrown upon the localities by the new maps.

* "Sinai and Palestine," 136.

Part IV.

THE HIGHLANDS OF WESTERN PALESTINE.

The Streams or Watercourses, together with the Waterpartings of a country, form the primary foundation of its delineation, and the proper basis of its study. The Plains require the next consideration, for they form the threshold of the mountains, and often determine their limits. The intricacies of the relief, and the complicated forms of the Elevated Ground, are thus besieged by regular approaches, leading up to and defining the base of the outer walls, and penetrating the interior by exact drainage lines, which, in order to ǀfulfil their functions, should not only define horizontal direction, but should also express by altitude, the vertical changes which they undergo.

The next step is the elimination from the highland mass of its Culminating Summits and the Ranges which are crowned by them. It is necessary to consider not only those ranges which form waterpartings, but also those which are intersected by rivers and watercourses, and make the edges of plateaus, often indeed, displaying a regularity, an altitude, and important consequences to nature, and particularly to mankind, transcending in effect the waterparting ranges.

To eliminate from the tangled maze of the western highlands of Palestine, the main ranges and groups that form a key to the whole mass, is the object of the following notes.

The rivers which define the limits of the survey under review, are singularly adapted to palliate the disadvantage of considering only an intermediate portion of the mountain system that extends in three directions beyond those bounds. The Kasmîyeh on the north is indeed a curt divider, as the name implies, separating the white and lofty Lebanon from the humbler summits that are its continuation southward through Galilee, if not also through Samaria, Judea, and onwards.

So also the Jordan cuts off the western highlands by a well-defined line from the eastern portions of the same general mass. But still it is necessary in treating on a separated portion to bear in mind its proper adjuncts.

Before the present Survey, a considerable number of observations for altitude had been made by various travellers. These were carefully collected, arranged, and more or less critically examined by Lieutenant Van de Velde, in his "Memoir to accompany the Map of the Holy Land."* Additions were afterwards made to the list, in a pamphlet by the same author, entitled "Notes on the Map of the Holy Land."†

The map of the Palestine Exploration Fund includes a large number of hypsometrical observations, and this series has the great advantage of having been taken on a uniform system, and by trained observers throughout, an advantage that is fully displayed by a comparison of the discrepancies exhibited in Van de Velde's lists. It cannot, however, be justly concluded that the new series supplies all that is wanted by the hypsometrical student. The observations appear to have been made casually rather than systematically. It would have been possible with a due regard to the horizontal continuity of the vertical development of the ground, especially with reference to leading features, to have contributed much more to an intelligible apprehension of the main factors of the relief, without incurring the labour of contouring. These remarks are consequent upon an experience of the difficulties arising from the occasional want of observations for altitude,

* Published by J. Perthes, Gotha, 1858.
† Published also at Gotha, 1865.

THE MOUNTAINS OF UPPER GALILEE. 177

in the following attempt to elucidate the orography of Western Palestine on the basis of the new survey. The defects are not organic, and may be supplied hereafter in this instance, and provided for in future operations.

THE MOUNTAINS OF UPPER GALILEE.

This mass of highland is bounded on the north by the Kasimîyeh River. The eastern and western boundaries are the Jordan and the Mediterranean Sea; the southern boundary is the foot of a Range which begins opposite Acre and runs eastward to the Jordan. From these bases, which will be more distinctly defined, the slopes ascend with much variety of feature, to culminating ranges on the south, the east, and west; enclosing an extensive plateau, broadest and highest on the south, and contracting to a narrow neck on the northeast.

The Southern Range.

The southern boundary is formed by a range of mountains commencing on the west in the Plain of Acre on the east of the city, and running eastward to the Jordan, below the Jisr Benât Y'akûb. It is distinguished by Kurn el Hennawy (alt. 1,872 feet); Neby Heider (alt. 3,440 feet); Jebelet el Arûs (alt. 3,520 feet); Es Semunîeh (alt. 2,235 feet); Safed (alt. 2,750 feet); Jebel Kan'an (alt. 2,761 feet). From the last-named point, the range descends to the gorge of the Jordan between Jisr Benât Y'akûb and the Sea of Galilee, where the river descends from 43 feet to 682 feet below sea level.

The summit of this southern range is thus easily defined by means of the observations for altitude made by the survey and the concurrent remarks of travellers. The base of the range is not so distinctly made out, for want of altitudes at the junctions, and other notable features along the watercourses. In one point of view, the base of the range is traced from the Plain of Acre along the Wady el Waziyeh,

Wady esh Shaghûr, Wady el Khashab, and Wady en Nimr, including the Plain of Rameh. But the proper base of this part of the slope is undoubtedly the Wady el Halzun and Wady Shaib, up to Wady el Khashab and Wady en Nimr. The last leads eastward to the waterparting of the Jordan and the basin of Wady er Rubudiyeh, the head of which is crossed by the footing of the range along Wady Said, and a valley on the south of Kefr Anan. Thus the base of the range is traced to Wady Maktûl, a watercourse on the north of Kh. el Bellaneh, which falls into Wady 'Amûd, and by that stream the base is carried to the northern shore of the Sea of Galilee, and up to the influx of the Jordan. See p. 148.

It may be remarked by the reader who pursues this description with the map before him, that the Wady et Tawahin intersects the range on the west of Safed. This is quite a common occurrence in mountain ranges, and is merely an indentation of the mass, without interrupting the continuity of the total elevation. The great range of perpendicular cliffs along the western side of the Dead Sea, is constantly rifted from summit to base, and the chasms, sometimes only scoring the face of the cliff, often run far back into the mass without affecting the continuity of its general aspects. In the case of the Safed gorge, although it has indented the summit of the range to the depth of a thousand feet, it has still to drop 1,500 feet to reach the base line coming from the west, and then has to fall another 900 feet to reach the Sea of Galilee. The distance of Safed from the mouth of the wady is about nine miles.

The Eastern Range.

The mountain range which bounds Upper Galilee on the south is continued along the eastern face of the highland to the top of the map. Commencing at the south-east with Jebel Kan'an (alt. 2,761 feet), the next observations along the summit occur in a group around Delâta (alt. 2,740 feet). Northward, none are found either at Ras el Ahmar, or at Alma, or at Jebel el Ghabieh, or along the waterparting south of Kh. el Menarah (alt. 2,806 feet). Further north, observations are found at Jebel Hûnîn (alt. 2,951 feet), Odeitha (alt. 2,215 feet), and Neby 'Aûeidah (alt. 2,814 feet). On some future occasion it would be desirable to fix the culminating points on the west of the Merj Ayun, also around the curious quadrilateral basin which is deficient of outfall, and includes the villages of Meis. The height of the culminating point along that part of the waterparting range between Kades and Aitherûn is wanted; and also the altitudes that connect Delâta with Jebel el Ghabieh, where the range is intersected by the remarkable rocky chasm, that communicates between the upper and lower parts of the Hindaj Basin.

The slope by which this Eastern Range descends to its base is varied and interesting, but the facts are inadequately demonstrated. The altitude of the base line is fixed at the Sea of Galilee at 682·5 below sea level; it is also expressed near the Jisr Benât Y'akûb, where the river is 43 feet below sea level. But at the Huleh Lake, the altitude of the surface is not inserted on the maps, and the Memoirs on sheet iv, while reporting that the observations made the surface of the lake seven feet above the Mediterranean, add that "they were not very good." In the Huleh Plain there is an observation excellently placed at the junction of the Jordan with all of its affluents, before the river enters the Huleh Marsh. This altitude is 140 feet above sea level. Others are given at El Mansurah (alt. 245 feet); at Kh. Dufnah (alt. 390 feet); at Tell el Kady (alt. 505 feet); at Kh. Dahr es Saghir (alt. 660 feet). This is the extent to which the altitudes partially elucidate the inclination of the base line of the eastern slope of Upper Galilee. It is an unfinished work. The course of the Jordan itself is cut short in the midst of its descent from the elevated hollow of Coele Syria to the much lower Plain of Huleh. Dr. Robinson called attention to the six successive terraces, by which the descent is made,[*] with a drop between each, generally not less than

[*] "Bib. Res.," iii, 391. "Phys. Geog. H. Land," 68.

50 feet, and sometimes more. The same authority also alludes to the curious way in which the channel of the Hasbany River forms a deep and narrow chasm along an upper terrace of the mountain side, instead of following the lowest ground. These distinct features are scarcely made out on the maps, and attention will no doubt be paid to them in the continuation of the Survey from its present limits to the east of the Jordan. It would be useful to define the base of the slopes, *or the line of lowest depression* in the upper part of the Jordan Valley, where the Hasbany departs from that line, and cuts its channel through the higher portion of the basaltic bank on the west of the valley. This base line seems to follow the Nahr el Leddan up to Wady en Nimr, and the latter wady till it forks at the foot of the slope which gives rise to 'Ain el Tineh. The line, doubtless, ascends the branch from the north, but the survey does not at present assist in carrying it farther. This part of the base line is of importance, as it divides the western slope of Mount Hermon, from the eastern slope descending from the Mediterranean waterparting, and it is an essential element in an attempt to understand the form of the ground in this interesting locality.

The slopes around the northern end of Huleh Plain may be resolved into three divisions. (1) Up to the northern termination of the Merj 'Ayûn, the slope from the Mediterranean waterparting may be considered to terminate in the stream which proceeds within the map from 'Ain Derderah, and runs to the Jordan as Nahr Bareighit. (2) The range on the east of Merj 'Ayun, dividing that plain from the Hasbany, may be conveniently regarded as the summit of the slope, along the eastern side of which, the channel of the Hasbany is cut, on its way to the Huleh Plain ; the base being found in a line along Wady Leddan and Wady Nimr to the Hasbany before it passes on to the slope; (3) The slopes of Hermon, ending in the base line before mentioned.

It is the slope of the second division which Dr. Robinson

found divided into distinct terraces; and the following is an endeavour to trace them on the new map. (1) The uppermost or first terrace appears to extend from the summit range on the east of Merj 'Ayûn down to a narrow and parallel edge on the east of the Hasbany. This terrace was formerly known as Ard Serada, and it includes the village of that name, also Kh. Jammûl, the village el Ghajir, and the channel of the Hasbany. Judging from the height of Abl (alt. 1,024 feet) that place should be included, but there is no clue to the southern edge. (2) The second terrace extends to Wady en Nimr. The direction of the first and second terraces, is from north-east to south-west, till the Hasbany bends due south near Kh. Jammûl, when the terraces take the same course. The third terrace shows a tendency to a more easterly front, and the remainder run east and west. (3) The third terrace is east of the base line formed by Wady Nimr, and at the foot of Mount Hermon. (4) The fourth excludes Tell el Kady, which is said to rise about 40 feet above the edge of this terrace and to drop about 80 feet to its southern base in the terrace below. The edge of the fourth terrace appears to run eastward as far as Tell Illa at the foot of Banias. (5) The fifth terrace follows the River Banias as it curves westward from Tell Illa, and leaves that river where it bends more to the south; the edge of the terrace then stretches across westward to Wady el 'Asl and Wady el Leddan. It follows the Leddan below Kh. Dufnah (alt. 390 feet), and crosses a stream from the second terrace, and the River Hasbany, between Kh. el Heit and Kh. es Sanbariyeh, west of the Hasbany; the edge of the fifth terrace seems to be defined by the course of an aqueduct, but in the absence of any altitude corresponding to that of Kh. Dufnah, there is nothing to follow. (6) The edge of the sixth terrace is probably found on the east of the River Banias, which the Survey barely touches. On the west of the river it seems to be indicated by an altitude of 302 feet at Baiket Francis, and follows the river to el Mansurah (alt. 245 feet), then stretches westward to a mill between the Leddan and the Hasbany, and pursues the same direction towards the Nahr Bareighit, above the village of Zuk et Tahta. Two or three altitudes westward of el Mansurah and corresponding with it, would have helped this attempt.

The structure of this slope is an orographical curiosity, and the notice it has received from such an eminent and exemplary observer as Dr. Robinson, makes it necessary in a Survey like the present, to meet it distinctly, either by adequate delineation or positive disavowal.

The connection of this slope with that between the Mediterranean waterparting and the Nahr Bareighit should also be elucidated. The following is a case in point. From Jebel Hunin (alt. 2,951 feet) the waterparting runs to the north-east, and bears on its summit the village and ruined castle of Hunin. Half a mile beyond Hunin, the waterparting takes a northerly direction, but a spur continues north-easterly to Nahr Bareighit, and has upon its brow the ruin of el Kuneiseh (alt. 1,064 feet). On the opposite bank of the river, in a more northerly direction, is Abl (alt. 1,074 feet). It seems probable that el Ghajir on the Hasbany is about the same height, but there is no observation. One or two altitudes near this level, on the east of Abl, would enable a hypsometrical connection to be traced between Jebel Hunin, Abl, and Banias (alt. 1,080 feet).

Recurring to Jebel Hunin, the waterparting on the north of that summit, *appears* to have its highest points at Odeitha (alt. 2,215 feet), and at Neby 'Aûeidah (alt. 2,814 feet). For rather more than three miles south of Jebel Hunin, the slope from the waterparting to the Huleh Plain, is rapid and sometimes rocky and precipitous, with an average fall of one in two. This aspect terminates at the distance mentioned, in a summit without a special name, but connected with Jebel esh Shakarah, and having five or six spurs diverging from it in all directions. The altitude is unknown. The waterparting which has reached this point from the north, here bends abruptly to the west, for about half a mile, then south-west and again westerly, to reach Neby Muheibib. This westerly diversion of the Jordan or Mediterranean waterparting is the southern limit of the isolated basins of Meis and Tufeh, and the northern limit of a great recession to the westward, both of the waterparting and of the summit of the eastern slope also, which run together from Neby Muheibib only as far south as Deir el Ghabieh. From the last place the waterparting again turns abruptly westward as far as Jebel Marûn, but the Eastern Range continues southward to Delâta (alt. 2,740 feet).

Returning now to the northern end of this Eastern Range, it will be found to have a parallel range on the west, which runs from Jebel Hunin to the Kasimîyeh River, and is in the same line with the range on which the great castle of Kulat esh Shukif or Belfort (alt. 2,345 feet) is situated. The valley between the waterparting and the parallel range on the west, is now called Wady 'Aizakaneh, instead of Wady Hunin, as on former maps. The Wady 'Aizakaneh is in the same line with the Kasimîyeh before it turns to the west, though sloping in an opposite direction. At their confluence occurs the rectangular bend of the Kasimîyeh, which diverts that river from a southerly to a westerly course, to become the northern base of the highland of Upper Galilee. The want of an altitude at this confluence, leaves the inclination of this northern base in obscurity, and prevents any comparison between the relative levels of the eastern and western bases of this portion of the Mediterranean waterparting.

The Wady 'Aizakaneh, although of small extent, becomes of further interest in connection with a succession of other features along the summits and slopes of the Eastern Range. That range has been seen to bend suddenly to the west as far as Neby Muheibib, when it turns again to the south. This inner or western parallel range will be found to be in continuation northward with the range on the west of Wady 'Aizakaneh. Starting in that direction from Neby Muheibib, this range forms the western boundary of the isolated basins of Meis and Tufeh; the northern limit of which begins at

Ras edh Dhahr and runs on through Jebel Husein, and Dahruj, to the Eastern Range, at a point rather more than half a mile on the south of Kh. el Menârah (alt. 2,805 feet). Meis is a favourite camping place for travellers, but none appear to have noticed its hydrographic isolation. From Ras edh Dhahr, the western range is intersected by Wady el Jemel and runs on to Jebel er Rueis. The Wady el Jemel drains a plateau on the north of the basin of Meis, and but for this outlet the plateau would form a similarly isolated basin. Beyond Jebel er Rueis, the range from the south, appears to meet the western range from the north, in Jebel Hunin; but when the height of Rubb Thelathin (alt. 2,292 feet) is compared with that of Merkebeh (alt. 2,290 feet), and of el Hola (alt. 2,470 feet), it may be inferred that the continuity of the elevation between Rubb Thelathin, Merkebeh, el Hola, and Ras edh Dhahr, calls for an accentuation of the hill drawing expressing those indications better than that of the present map. At all events, the continuity of the western range from Kulat esh Shukif to Deir el Ghabieh is made out, whether it joins Jebel Hunin or runs through Merkebeh and Hola.

The interval, properly so called, which has been traced from the Litany and Wady 'Aizakaneh on the north, to the basin of Meis on the south, does not stop at Meis, but is prolonged further south in the well known terrace or plateau of Kades (alt. 1,587 feet), the only altitude that is found from one end to the other of this intramontane tract. The eastern edge of the plateau of Kades is the prolongation of the lower part only of the Eastern Range in the line maintained by the upper part also, up to the northern end of the plateau. The plateau of Kades occupies the space caused by the westward recession of the waterparting along with the upper part of the Eastern Range. But it should not be overlooked that between Kades and the waterparting lies a still higher terrace, running parallel with the waterparting and above Kades, from el Malkîyeh to Belîdeh, Kh. el Maserah, and up the Khallet Ghazaleh to the unnamed summit, where the westerly recess of the Eastern Range begins. This is not a new feature, but it has never been delineated before with the precision of the New Survey.* These terraces of Malkîyeh, Belîdeh, and Kades, are drained into the Huleh Marsh by the steep descent of Wady 'Arûs. But the plateau itself seems to have a more gradual passage to the base of the mountain by the Ard el Dawamin, which forms a step downwards to Ard el Kheit and the shores of Lake Huleh. This step is included in a small basin, the western limit of which is in a line with the eastern edge of Kades plateau. There is indeed in this succession of natural features, confined between parallel ranges, a prolonged line of elevated lateral communication, extending from the Ard el Kheit, through the Kades plateau, Meis and 'Aizakaneh, to the Kasimîyeh; and if the Survey extended further north the line might be pursued to the fork of Wady Jermuk with the Litany, and so continued either by Coele Syria or by the Wady Jermuk, the Upper Zaherany, and the Upper Auwaly to the latitude of Beirut.

There is one more occasion to recur again to Wady 'Aizakaneh. Parallel to that valley, but much more extensively developed both in length and breadth, is another affluent of the Kasimîyeh, coming from the south and rising at Marûn er Ras (alt. 3,083 feet), in the midst of the higher plateau of Upper Galilee. Its principal channel is Wady Selukieh, the name of Wady Hajeir which formerly assumed that distinction, being now applied to a tributary. Although Wady Selukieh belongs to the northern slope, and will have to be considered in connection with that part of the subject, it is also related to a portion of the eastern slope, in connection with the upper part of Wady Hindaj, which enters Huleh Lake on the south of Ard ed Dawamin.

The Wady Hindaj marks a fresh and striking change in the features of the Eastern Range. The recess distinguished by the Plateau of Kades is backed by the Eastern Range and waterparting between Neby Muheibib and Jebel el Ghabieh.

* Robinson, "Bib. Res.," iii. 367.

The latter point is on the northern edge of the Wady Hindaj basin, the limits of which require attention, as they shed a light on features of the range that have been already noticed, as well as upon those that begin here. Beginning at Lake Huleh, the northern edge of the basin runs in a south-westerly direction to a range of hills that forms a prolongation of the original line of the eastern range. The boundary runs northward along this range to the south-eastern corner of the Kades plateau, where it turns westward along the southern edge of the plateau to Deir el Ghabieh. From this point the waterparting of the Mediterranean and Jordan and of this basin makes a further departure westward, as far as Jebel Marûn (alt. 3,050 feet); while the Eastern Range appears to cross the basin, from the same point southward, in the direction of Delâta (alt. 2,740 feet). A large basaltic dyke observed by Dr. Tristram between Delata and Alma falls in the line of the range. A large patch of basalt farther north seems to be Jebel Gabieh in the same line. Perhaps the smaller dyke to the eastward is the ridge dividing the plateau of el Malkîyeh from the Merj Kades.* At Jebel Marûn the waterparting bends abruptly to the south-south-west as far as a point on the range south-east of Jebel Adather (alt. 3,300 feet). There it turns south-east to Jebel Jurmuk (alt. 3,934 feet), the highest point in Galilee; then it goes off on a general course to the north-east as far as Alma, where the Eastern Range crosses the basin and makes the waterparting deflect to the south-east in descending to the base of the range in the Ard el Kheit and the Huleh Lake. The wady also descends the range by a rocky and precipitous chasm, in a parallel direction to the waterparting; but on reaching the plain, the stream bends suddenly to the eastward with a little northing, and so crosses the Ard el Kheit to Lake Huleh.

The elevated part of the Hindaj basin extends so far west as to intercept the heads of Wady Selukieh running north to the Kasimîyeh, and of Wady 'Amûd running south by Safed to the Sea of Galilee. The 'Amûd, the Upper

* Tristram's "Land of Israel," 377.

Hindaj and the Selukieh, with the bridge of K'ak'aiyeh across the Kasimîyeh, thus become a fully developed line of lateral communication connecting the Sea of Galilee and the Lower Jordan with the Sidonian coast. Thus is explained the intimate connection between Sidon and the Upper Jordan. Josephus calls the Huleh "The great Sidonian Plain."* This is the relation of the Wady Selukieh with the eastern slope to which reference was made partially in the eleventh page of this work.

All this division of the eastern slope, excepting its westernmost portion along the Mediterranean waterparting, but including its high plateau as well as the range which crosses the slope, and the descent to the base in the Ard el Kheit, is characterised by rocky precipices and deep chasms. Its base is found only falling back a little from the line which it has maintained from the extremity of the Survey up to the basin of the Hindaj. The ascent from this base no doubt culminates in the Eastern Range passing through Delâta and Alma, across the Hindaj chasm to the range on the west of the Plain of Kades. The descent thus limited is without the intermediate range or the plateaus that characterise this eastern mountain further north, but chasms or precipices crossing the slope, are substituted very prominently.

South of the Hindaj basin, the eastern slope is much restricted in its westerly extent, by the interposition of the 'Amûd or Safed basin. At Ras el Ahmar the waterparting is somewhat more westerly than at Deir el Ghabieh on the north, but the highest ground is considered to be further east, in a line between Delâta and Alma, where Dr. Tristram's basaltic dyke occurs. At the southern termination in Jebel Kan'an on the east of Safed, the Eastern Range has quite returned to its former meridian, and the slope is eased off by a terrace containing the villages of Jû'auneh, Feram, el Mughar (alt. 1,670 feet), and Kabbâ'ah (alt. 1,745 feet), also by spurs advanced into the Ard el Kheit, and raising its level at Kh. el Loziyeh to 487 feet, and at 'Ayun el Wakkas to 365 feet.

* Ant. v, 3, 1. Judges xviii, 7, 28.

It is yet too soon to dismiss the features connected with the Hindaj. It is the only part of the eastern slope of Upper Galilee that comes in contact with the Mediterranean waterparting, where the latter throws off its channels directly to the westward. On the south of the Hindaj, the plateau of Safed also abuts on the Mediterranean slope, but that plateau is not on the eastern, but on the southern slope of Upper Galilee. The plateau of the ' Selukieh which follows that of the Hindaj on the north, belongs to the northern slope.

Another remark on the Hindaj may be made. North of the Hindaj, the waterparting of the Mediterranean makes a succession of steps to the eastward, so as to become east of the Litany and the Lebanon. The western limit of the Hindaj is the commencement of the meridional direction, along which the Mediterraneo-Jordan waterparting zigzags for the rest of the Survey. Its most westerly point near Sasa is the first of a series of projections to the west, alternating with recesses to the east, which mark the progress of the great parting southward. The origin of this configuration will probably engage the attention of future observers.

The direct *orographic* continuation of the main waterparting becomes disconnected from its hydrographic accompaniment, to the north of the Hindaj, and pursues a line on the west, instead of on the east of the Selukieh; and reaches the Kasimîyeh in a bend of the river northward, below the confluence of the Selukieh, and the K'ak'eiyeh Bridge. It is a line with successive oscillations, that point to a continuous series of small elevated plateaus alternating from one basin to another, and which will be noticed hereafter.

The parallelism in the orography of Upper Galilee, which has been already remarked, extends beyond the limits of the eastern slope and its ranges, and includes a part of the western slope. But as the extraneous parallel has a direction in common with those already noticed, and has a dividing line also in common with the Safed, Hindaj, and Selukieh series, it seems desirable to deal with it at once, and to make it the

THE WESTERN SLOPE. 189

beginning of the examination of the western part of Upper Galilee.

The Western Slope and its Upper Plateaus.

The line of waterparting heights that forms the western boundary of the valleys of Safed or 'Amûd, Hindaj, and Selukieh, divides that series of valleys from another on the west, that in some respects presents similar facilities of lateral communication through the highland, between north and south.

Eastward of the roadway carried over the plains, and the occasional headlands along the coast, passage in a parallel direction is obstructed for a considerable distance inland, by the constant succession of deep and narrow valleys, and steep ridges descending from the interior. In the direct distance of about 30 miles between the River Kasimîyeh and Acre, there are no less than 30 wadys or rivers and channels which have distinct outlets into the sea, and which have to be crossed in travelling northward or southward. To this number of main channels must be added the valleys of their numerous affluents, and the ridges between them, running for the most part in a more or less parallel direction. These obstructions in the hills to communication parallel with the maritime plain cover a belt of country exceeding a dozen or fifteen miles in breadth. Within that belt are found the sources of all the wadys contributing to the 30 outfalls before mentioned, except three. These three are the Hubeishîyeh, the Ezzîyeh, and the Kûrn. They pass through the furrowed belt to the sea, in deep and narrow gorges like the rest, but in the higher ground beyond, they spread out into consecutive plateaus, and overlap the others, their main streams assuming more or less of a meridional direction; while the branches running north and south, greatly facilitate communication.

The plateaus formed by the upper parts of the Hubeishîyeh, Ezzîyeh, and Kûrn basins, include the western division of the highland of Upper Galilee; the eastern division being embraced mainly by the plateaus of Safed and Selukieh, supplemented by the smaller series of 'Aizakaneh, Meis, and

190   THE MOUNTAINS OF UPPER GALILEE.

Kades. The edges of the western highland form the Western Range which will now be made out from the New Survey.

The Western Range.

The Southern Range will·be used as the starting-point of the Western, as it was of the Eastern Range. Although no intermediate altitude is found on the Southern Range between Neby Heider (alt. 3,440 feet) and Kurn Hennawy (alt. 1,872 feet), yet the height of Kisra (alt. 2,520 feet) on the north of the Southern Range, indicates the junction of the Kisra spur with the Southern Range, as the probable point of departure for the culminating summits of the western slope.

A cue to the course of the Western Range is found in the western-most observations rising above 2,000 feet. That guidance leads from the short spur dominated by Kisra, north-westward to Kh. Jubb Ruheij (alt. 2,320 feet), near Yanûh (alt. 2,200 feet). North of Jubb Ruheij is Kh. ed Dubsheh (alt. 2,050 feet) near Teirshiha (alt. 1,810 feet). About a mile further north is Malia (alt. 1,800 feet), and about the same distance beyond, this line of heights is intersected by the rocky gorge of Wady el Kurn. Two miles north of the gorge, the line of heights is taken up by Tell Belat (alt. 2,020 feet) in the secondary basin of Wady Kerkera.

Before leaving the line of the western heights on the south of the Wady el Kurn, it should be observed that it corresponds very closely with the waterparting which divides the upper expansion of the Kurn basin, from the heads of a series of five minor basins, that have their outfalls into the sea between Wady el Kurn and Nahr N'amein. On the inner, upper, and eastern side of this part of the Western Range, the principal watercourses are parallel to this range; while on the outer, lower, and western side the watercourses rising in the range, pass off towards the sea, at right angles to it. The Western Range crosses the Wady el Kurn, where the gorge running from south to north, takes a westerly course, and descends to 495 feet above the sea, at the foot of Kulat el Kurein, the remains of the Crusaders' castle of Mont Fort, which commanded the pass. The altitude of the river before it crosses the mountain, compared with its altitude west of the range, would probably indicate a considerable fall, and throw light on the height of the plateau at its lowest point.

Three miles about north-north-east of Tell Belat, the prolongation of the Western Range is defined by Khurbet Belat (alt. 2,467 feet). The eastwardly direction corresponds with the change in the course of the coast line on the north of Ras en Nakurah. From the northern bank of Wady el Kurn to Khurbet Belat, the range crosses the secondary basin of Wady Kerkera. It is a spur from Kh. Belat, which terminates at the sea in the headland of Ras en Nakurah or the ladder of Tyre; and by the eastern end of that spur the Western Range reaches the summit of the mountain.

The Khurbet Belat is remarkable for the ruins on its summit, and also for one of the grandest panoramic views in the country. It rises on the southern edge of the Ezzîyeh Basin, and besides throwing off a spur to Ras en Nakura, it sends off another spur to Ras el Abiad or White Cape. The western range passes from it across the Ezzîyeh Basin, to Ras el Bedendy (alt. 2,215 feet), which has the village of Yater close on the west, with an altitude already fallen to 1,589 feet.* About a mile to the north, the range is found at Ras Umm Kabr (alt. 2,341 feet); from whence it passes east to Harîs (alt. 2,343 feet), then north-east to Kh. el Yadhun (alt. 2,612 feet), Jebel Jumleh (alt. 2,625 feet), and Kh. Selem (alt. 2,219 feet). At the last point the Western Range terminates, and the northern limit of the plateau forms its continuation.

It will be observed that where the range passes from Kh. Belat across the Ezzîyeh basin to Ras el Bedendy, the basin contracts from the expansions of the upper part, both on the north and on the south; and the principal watercourse also bends from a meridional direction suddenly to the westward by a very tortuous channel. Below the mountain, the gorge

* Robinson's "Bib. Res." III, 61.

becomes a deep and rocky chasm, with branches of the same character.

Between Ras el Bedendy and Jebel Jumleh, the range crosses the Hubeishiyeh basin, and sends off from Kh. el Yadhun, the spur which has upon it the Crusaders' castle of Tibnin or Toron. This spur lies between two curious divisions of the Hubeishiyeh basin. The uppermost is the head of the basin. It is nearly four miles long by two miles broad, and slopes from the south-east towards the base of the Tibnin spur, along which a wady runs to the north-east, and receives several streams from the south-eastern slope. At the north-east corner, the wady discharges the drainage of this enclosed plateau, by passing round the end of the Tibnin spur, and then entering the south-east corner of a similarly enclosed plateau, it skirts the spur along its north-western base, in the reverse direction to its previous course on the other side. The wady is turned abruptly from the spur in a northerly direction by the range between Yadhun and Jumleh, which crosses the eastern part of this plateau obliquely. The wady intersects the range, and being then diverted to the westward by a parallel range, takes the name of Wady el Ma, and reaches the north-western corner of the plateau, where it receives a tributary from the higher western range, and leaves the plateau, proceeding on a north-westerly course to the sea. In this direction it keeps close along the southern boundary of the basin, and receives all its affluents from the triangular slope between it, the north-western side of the hills enclosing the Wady el Ma, and the spur descending from Jebel Jumleh through Teir Zinbeh, Silah, and Mahrakah. This spur separates the northern and southern divisions of the Hubeishiyeh basin. The main channel of the northern division of this basin, called Wady Humraniyeh towards its outlet, also hugs its southern limit in the same way, and receives nearly all its affluents from the north, where the waterparting divides the basin from that of the Kasimiyeh. There are three of these affluents, rising in succession, and running at first in the same line, parallel with the main stream, until each of them bends at a right angle, and descends the slope to its confluence. They are perfect examples on a small scale of the lateral communication which is seldom wanting among hills or mountains.

Looking back southward, over the lower part of the western slope of Upper Galilee, it will be found to present three distinct divisions, in consequence of the elevation of the central division which lies between Wady Ezzíyeh on the north, and Wady Kerkera on the south. Between these basements, the spurs descend to the sea from the Western Range, where it rises up to the summits of Kh. Belat (alt. 2,467 feet), and Tell Belat (alt. 2,020 feet). It is however the former alone that is the centre of the elevated spurs that distinguish this central tract. To Khurbet Belat may be distinctly traced the waterparting ridge which has on its summit Birket er Rahrâh (alt. 1,865 feet), and El Mejdel (alt. 1,375 feet). Near the latter place, the waterparting forks and divides the small Wady el Mansur from the great Wady Ezzíyeh, and the minor Wady Shema. As el Mejdel is only two miles from the plain of Tyre, and but three miles from the sea, the descents into the plain of the spurs on either side of Wady el Mansur, must be somewhat abrupt. Another spur is defined by the village of Shihin, and Kûlat Shema (alt. 1,255 feet). It spreads out to a mile in width, west of the Kulat (castle), and concludes in the line of cliffs that has the Ras el Abiad (White Cape) at its southern end, and terminates the Tyrian Plain. About a mile nearer the sea than Castle Shema is Kh. Kermith (alt. 1,290 feet) on another spur apparently free from cliffs. Further south, Kh. Umm 'Ofeiyeh (alt. 900 feet) is only a mile from the sea. The spur on which it stands comes down from Kh. Belat through Shihin, El Jubbein, and Teir Harfa. The last spur of this central division is the most important of the series, and passes down from Kh. Belat, under the names of el Menarah, Tell el Kishk, and Jebel Mushakkah; and finally reaches the sea at the famous headland of Ras en Nakurah, the Hewn Cape, and Ladder of the Tyrians, or Scala Tyriorum, of the ancients.

Jebel el Mushakkah may be used as the name of the whole of this range from Ras en Nakurah, to the foot of Kh. Belat, near Birket Risheh. Its northern base may be found in Wady Zemzem, Wady es Serawat, and Wady es Zerka. Its southern base is Wady Abu Muhammed and Wady Kerkera, as far as the edge of the Plain of Acre, when the base line is carried northward to Wady el Kutayeh, and along the north side of the gardens of el Basseh and el Musheirefeh to the sea. The altitude of 1,192 feet, is given to the trigonometrical station on the summit of the mountain between el Basseh and Lebbuna, about two miles from the sea. An altitude of 225 feet at Ras en Nakura, is probably applied to the roadway cut in the face of the cliff. The slope towards the plain of Acre is very rapid, perhaps rugged, but apparently devoid of cliffs. East of the plain, the southern slope expands to the south, while the ridge recedes a little to the north, so as to produce a space of nearly two miles in width between the ridge at Tell el Kishk and the base in Wady Kerkera. This space forms a terrace or plateau drained by Wady ed Delem, and bounded on the south by a succession of bold cliffs with ruins above them, among which is Kh. Idmith (altitude 1,810 feet) with others which have been described by Mons. Guerin.* The northern slope spreads out towards the receding shore line, it is more varied, has many interesting ruined sites, and also the existing villages of Alma esh Sh'aub (altitude 1,360 feet), Lebbuna and en Nakura (altitude 221 feet). The slope from Jebel el Mushakkah to the sea is limited by a spur from Alma to en Nakura, the northern slope of which gives rise to its own watercourses, partly contributing to the lower course of Wady ez Zerka. This wady was formerly called by Robinson and others, Wady Hamûl, and the Ain Hamûl is still connected with it on the new map. Another spur contributing to the character of this slope, proceeds from the west end of that part of the range called el Menarah, and follows the Wady Zerka on the south as far as Tell esh Shatin.

* Galilée, II, ch. lxxviii.

From the east end of el Menarah, the remainder of the slope descends to the base of the mountain in Wady Zemzem and Wady es Serawat. The archæology of this division has attracted the attention of MM. de Saulcy, de Vogué, Renan, and Guerin.* Dr. Thomson notices the coast route.†

The divisions of the western slope on the north and south of the central portion, descend to the plains of Tyre and Acre respectively. The slope of the northern division, in descending from the Western Range, appears to present features that separate the upper from the lower part. It will be found that all the larger wadys at first run northward, with more or less westing, and that they bend suddenly to the west, at points in their courses that follow in a continuous line, parallel with the coast. On the outer side of the same line, minor streams, both independent and tributary, take their rise. The conclusion is that the slope makes a sudden drop along this line, which forms a kind of retaining wall to the upper part of the streams, and is broken through by the streams in descending to the lower part of the slope. It may also point to some geological fact that has hitherto escaped notice.

In the southern division of the western slope, the course of the main valleys is scarcely at all northerly, and if an intermediate escarpment exists, it must be looked for in another form. Such a point may be noticed as Yerka (alt. 1,200 feet), in comparison with Kefr Yasif (alt. 279 feet), on the same spur. Due north of Yerka, on Wady el Kurn, is found Kulat el Kurein (alt. 1,055 feet), and Kh. Menhatah (alt. 1,405 feet). Perhaps an examination might be fruitful along the connecting line, including additional altitudes at or near Ras Kelbân, Kh. el Habai, Kulat Jiddin, and Kh. Zuweinita.

---

* Guerin, "Galilée," II, chaps. lxxvi, lxxvii.
† "Land and Book," 302.

The Northern Range.

Between Khurbet Selem and the Kasimîyeh, which is the northern limit of Upper Galilee, it is necessary to look due east for the next altitude of 2,000 feet; and one is found at Merkebeh (alt. 2,290 feet), on the inner of the two parallel Eastern ranges. The actual junction of the Northern with the Eastern Range, is no doubt further south, and due west of Hunin, which will become evident in tracing the range from Kh. Selem eastward across the valley. Kh. Selem overhangs the deep and precipitous valley of el Hajeir, and the continuation of the summit of the range is found on the opposite bank, in the village of Suwaneh (alt. 1,766 feet), from whence it takes a zigzag course to Mejdel Islim (alt. 1,910 feet). The short slope of this part of the range is towards the south, where it has for its base the Khallet el Dalieh; while the long outer slope, extends through Tulin and Abrika, to the great bend of the Selukieh, where it joins the Hajeir, and runs on to the Kasimîyeh. The Hajeir also gives its name to the main wady below the junction, although it is less than half the length of the Selukieh. East of Mejdel Islim, the range is intersected by Wady Selukieh, and on the other side of the wady, it is proposed to trace it along the ridge on the north of Wady el Beiyad, up to the Eastern Range. The slope of this part is short to the south, while the northern slope is longer and extends through Tallûsah to the bend of Wady Selukieh. Below the Northern Range, the heights descend to a general level of about 1,500 feet, and drop abruptly by rocky precipices to the lowest wadys which are about 600 feet, at the junction of the Hajeir with the Kasimîyeh. There is an observation at the junction of W. Bureik with W. Hajeir (alt. 635 feet). Above the range the country rises up to 3,000 feet.

The Interior of the Plateau of Upper Galilee.

It is now time to take a glance at the interior of the high plateau of Upper Galilee, within the ranges that form its outer limits.

From Jebel Mugherat Shehâb, on the east of Jebelet el 'Arûs (alt. 3,520 feet), the latter being the highest point observed on the Southern Range, another range runs north-westerly, crossing the plateau obliquely to Kh. Belat in the Western Range. This line of heights has the basins of Kurn and Kerkera on the south-west and those of Safed, Hindaj, and Ezziyeh on the north-east. It is traced through Kh. Zebûd (alt. 3,200 feet); Jebel Jurmuk (alt. 3,934 feet), the highest point in Galilee; and Jebel 'Adâther (alt. 3,300 feet). Northward the range runs between Katamûn and Semmûkhieh, both of which exceed 2,400 feet, but between Jebel 'Adâther and Kh. Belat no observations are found along this line. This range may be called the Jermuk Range, after its culminating summit.

Another line of heights runs obliquely from Deir el Ghabieh on the Eastern Range, to Kh. el Yadhun (alt. 2,512 feet) on the Western Range. This line exceeds 3,000 feet at Jebel Marûn, and it is 2,675 feet at Kh. Shelabûn. It divides the basins of Hindaj and Ezzîyeh, from those of Selukieh and Hubeishiyeh. This line may be called the Marûn Range. According to abundant observations in the upper part of the Ezziyeh basin, the principal watercourses average about 1,800 feet, and the hills about 2,475 feet above the sea. The observations in the upper part of the Kurn basin are very few, and not sufficient for any purpose.

The plateau on the south-west of the Jermuk Range, consists of the two main valleys of the Upper Kurn, and the heads of the Kerkera. The latter, and one branch of the Kurn, receive the south-western drainage of the Jurmuk Range. This branch of the Kurn is situated, according to Dr. Robinson,[*] in a " deep and vast valley ; " and the aspect of the

[*] " Bib. Res.," III, 75, 76, 77.

country viewed from Beit Jenn, at the head of the valley in April, is said by the same authority to be "bald, barren, and desolate, in the highest degree." The other branch of the Kurn drains the Bukeiah, which signifies a hollow between mountains, here forming a well-cultivated plain, and including among its population, who are chiefly Druzes, a few Jews, who are said to be the only Jews in Palestine engaged in agriculture, and who claim descent from families settled in this remote highland from time immemorial. This branch of the Kurn drains the interior slope of the Western Range. Between Teirshiha and Suhmata, the plain or hollow terminates in a deep and rocky gorge, at the outlet of which the two branches of the Upper Kurn unite, before descending westward to the sea, through the deep and rocky bottom of the narrow neck, into which the basin contracts on the western slope.

The north-eastern side of the Jurmuk Range seems to be characterised by its plains, forming fine tracts of cultivated land with plenty of pasture, woodlands, and orchards. Picturesque hills and valleys, and rocky glens, villages and vestiges of antiquity; horses, cattle, and camels, sheep and goats, mules and asses, cats and dogs, poultry and game, birds and beasts of prey, contribute to the scene.

The plains spread out over the upper parts of the basins of el 'Amûd, Wakkas, Hindaj, and Ezzîyeh, and they extend from Meiron to el Jish and Delâta, Alma, Salhah, Yarun, and Rumeish. This wide circuit surrounds a higher tier of more undulating ground, backed by the Jurmuk Range, and containing the villages of Kefr Birim and Sasa. On the north, the plains are bounded by the Marûn Range, which forms a waterparting summit to a broad expanse of deeply fissured and densely wooded highland, abutting against the Western Range, which is indeed its escarpment towards the sea. Aligned between north-west and south-east, the Marûn Range extends its spurs and valleys to the Upper Ezzîyeh on the south, and to the Upper Selukieh and the Eastern and Northern Ranges. Between Ras el Tireh, and Kh. el Yadhun, the range also contributes to the head of the Hubeishiyeh, which lies above the castle of Tibnîn. The portion of the Upper Ezzîyeh referred to, bears the names of Khallet el Mukeisibeh, Wady el Malek, and Wady el 'Ayûn, on the one-inch map; and this channel forms so far a common base for the Jurmuk and Marûn Ranges.

The Mountains of Lower Galilee.

The northern base of the mountains of Lower Galilee is of course identical with the southern base of the mountains of Upper Galilee, which has been already described and explained. It will be sufficient to repeat now that the northern base line runs from the Plain of Acre, along Wady el Halzun, Wady Shaib, Wady el Khashab, Wady en Nimr, Wady Said, Wady Maktûl, and Wady 'Amûd to the Sea of Galilee. The eastern limit is the Sea of Galilee and the Jordan. The western and southern limits are formed by the Plains of Acre and Esdraelon, and the Nahr Jalûd.

The difference in altitude between this region and Upper Galilee is relatively considerable; for while the latter nearly attains to a height of 4,000 feet, the hills of Lower Galilee never rise to 2,000 feet. The general features of the upland of Lower Galilee, are also very different from Upper Galilee. They present a succession of parallel ranges, divided by broad plains; the ranges running between east and west, with a slight bend towards the north.

The Northern or Shaghûr Range.

The northern range rises from the base line common to Upper and Lower Galilee, which has just been traced in detail. Its southern base is defined by the Wady el Melek; continued upward as Wady Khalladîyeh, to the Plain of Buttauf and thence along that great plain, and the southern foot of Jebel et Teiyarat, to Wady Sâd, and the sinuous, deep and rocky gorge by which the Wady el Hamâm reaches the Plain of Gennesaret and the Sea of Galilee, on the north of el Mejdel. About half a-mile above the junction of Wady Sâd, with Wady el Hamâm, the wady is at the level of the sea, and there the depression of the Jordan Basin commences in this locality. The central part of this upland gives its name to the District of esh Shaghûr; the western is in the District of Shefa 'Amr, and the eastern descends from esh Shaghûr, by the Wady er Rubudîyeh to the Plain of el Ghuweir or Gennesaret. The western part of these hills has been noticed on p. 124.

The summit of this upland is a plateau between two ranges, one of which is easily defined, for it is coincident with the waterparting between the N'amein and Rubudîyeh basins on the north, and the Mukutt'a and el Hamâm basins on the south. The Survey supplies no altitudes along this range, west of Jebel ed Deidebeh (alt. 1,781 feet). The altitude at Kh. Jefat (alt. 1,363 feet), is no doubt inferior to that of the range, which was observed again further east at Ras Kruman (alt. 1,817 feet) and at Ras Hazweh (alt. 1,781 feet). The range lies between Kh. Natef (alt. 635 feet) and Ailbun (alt. 515 feet), and it is doubtless higher. The range continues along Jebel et Teiyarat, and el Muntar, to War 'Atmeh, but no heights have been observed along this waterparting.

The other range did not escape the notice of that masterly observer, Dr. Robinson, and with his assistance chiefly, the new survey enables it to be distinctly traced. Beginning with Shefa 'Amr ('Omar), Dr. Robinson remarks that it is " on a ridge overlooking the plain."* The altitude on the map at 'Ain Shefa 'Amr, is 136 feet. But that must represent the edge of the plain at the foot of the ridge, for Captain Mansell, R.N., ascertained the height of Shefa 'Amr to be 533 feet.† The width of the plateau is here about two miles, or the distance of Shefa ' Amr from the waterparting ridge on the south, which is stated by M. V. Guerin to be 200 metres, or about 650 feet at Jebel Kharouba, a summit occupied by Saladin during his conflict with the Crusaders under Richard the Lion-hearted, and Philip Augustus.

The range is continued through 'Abellîn, two miles north-

---

* Rob. "Bib. Res." III, 103.
† Admiralty Chart.

east of Shefa' Amr. 'Abellîn is "perched upon a high and sharp hill," 526 feet above the sea, according to Captain Mansell, R.N.,* and "also looking over the plain." The range goes on to Ras Tumrah (alt. 1,150 feet) "on the top of the first ridge, affording a noble view," etc. Thence to M'âr (alt. 872 feet) "on the western brow of the mountains, overlooking the great plain along the coast." The next point is Jebel Khanzireh (alt. 1,320 feet), where the ridge overhangs its base on the deep Wady Shaib. On the east of this mountain, the range is intersected by the gorge of Wady el Jizair, beyond which it is represented by Jebel el Kummâneh, and Jebel Hazzûr, both dominating the base line which is here found in the Plain of Rameh. From Jebel Hazzûr it proceeds without interruption to Jebel el Bellaneh (alt. 1,150 feet), and terminates on the banks of Wady 'Amûd.

On the long plateau between these ranges is found the upper part of Wady 'Abellîn; Kaukab (alt. 1,330 feet); Suknin, the ancient Sogané (alt. 910 feet); the fertile Plain of 'Arrâbeh, the Castle of Deir Hanna (alt. 1,070 feet); the Plain of Selameh "covered with olive groves," and containing the ancient site of Selamis, a town fortified by Josephus.

The Toran Range.

The Jebel Toran commences at Rummâneh, considerably short of the western limits of the Shaghûr Mountains. But the eastern development of the range makes amends for this, by its extension along the western shores of the Sea of Galilee and the course of the Jordan, as far as the descent of Wady Fejjas from the Plain or Sahel el Ahma to the Ghôr. The northern base of the range coincides with the southern base of the Shaghûr Mountains, from the Plain of Buttauf eastward. The southern base is defined by the Wady Rummâneh and the Plain of Toran, passing eastward by the Hakûl el Mughârah to the north of Lubieh; whence it descends by Wady Shubbâbeh and Wady Fejjas, through the

* Admiralty Chart

Sahel or Plain of El Ahma, and the rocky gorge of Fejjas, to the Ghôr and River Jordan at Umm Junieh.

The range describes a bold curve parallel with that on the north, and it culminates in a single ridge throughout. Steep escarpments are displayed towards the Plain of Buttauf and the Sea of Galilee, with rocky cliffs towards el Mejdel, Tiberias, and the Fejjas gorge. The central part of the range expands into a broad plateau, around Hattin and Nimrin. This plateau extends between Jebel Toran on the west, and Hajaret en Nusâra on the east; with the Merj Hattin on the north, and the plain between Hattin and Lubieh on the south, where the decisive victory of Saladin over the Crusaders took place. The great plain of el Ahma lies on the south-east of this plateau, and is chiefly formed by the western slope of the range where it skirts the Sea of Galilee; as the slopes on the right bank of the base line ascend steeply to the summit of the next range.

The culminating points of the Toran Range are Jebel Toran (alt. 1,774 feet); Nimrin (alt. 1,110 feet); Kurn Hattin (alt. 1,038 feet); Hajaret en Nusâra (alt. 740 feet); Tell M'aun (alt. 715 feet); el Menarah (alt. 966 feet). To the height of Kurn Hattin and the summits which follow it, the depression of the Sea of Galilee (—682·5 feet), should be added, in order to represent the actual elevation of those points above their eastern base. Hajaret en Nusâra, or Stones of the Christians, is the reputed site of the miraculous feeding of the four thousand with seven loaves and a few fish. M. Guerin, in contending for this site, carefully distinguishes the preceding event from the feeding of the five thousand with five loaves and two fishes, which requires a locality on the eastern shore of the Sea of Galilee.* But a site on the east of the sea appears to be necessary for both.

The Nazareth Range.

This division of the Hills of Lower Galilee is at first sight, of a complicated character; but by strictly defining its

* Guerin, "Galilée," I, 185.

limits, with regard to its bases, and also to its line of culmination, it will be reduced to an intelligible form.

The northern base is, as a matter of course, identical with the southern base of the preceding range, from its eastern end on the Jordan, to its western end at the passage of Wady Rummâneh from the Plain of Toran into the Plain of Buttauf. From that point the more westerly extension of the Nazareth Hills, requires the base line to be continuous along the same channel to Wady el Khalladîyeh and Wady el Melek, up to the junction of the latter with the Mukutt'a.

The southern base proceeds on the east from the Jordan, up Wady el Bireh and Wady esh Sherrâr, to the south-western foot of Mount Tabor or Jebel et Tur, where it approaches the waterparting of the Mukutt'a; and the Wady then receives affluents from Iksal on the north-west, and from Jebel Duhy on the south. From this confluence the base is followed across the waterparting to the affluent of the Mukutt'a which descends by Bir el Hufiyin and Wady el Muweily to the Nahr el Mukutt'a, which it follows to its confluence with Wady el Melek, or so far as the hills extend into the Plain of Acre.

The main summits of the hills within these limits, may be traced from the Plain of Acre, by a spur which rises from the plain near el Harbaj, and which forms a portion of the waterparting between the tributaries of the Mukutt'a and those of its affluent, the Wady el Melek. This waterparting, which will be more fully elucidated hereafter. leads eastward to a summit (alt. 1,548 feet), about a mile on the west of Nazareth, and this is the first observation for altitude so far along the waterparting. Observations on either side seem to imply that on the west of Nazareth it does not descend below 700 feet, until its final decline into the Plain of Acre. The lowest depression, which forms a notable point in the further description of the range, will probably be found between the southern end of Wady el Khalladîyeh, and Zebdah (alt. 350 feet), in the Plain of Esdraelon, on the west of Semûnieh.

At Neby Sain, immediately on the north of Nazareth, the alt. is 1,602 feet. About a mile east of Nazareth, the waterparting between Mukutt'a and Melek joins the main waterparting between the Mediterranean and the Jordan, and it would be desirable to ascertain the altitude there.

The main range now follows the Mediterranean waterparting to Jebel es Sih (alt. 1,838 feet), and pursues it on a north-easterly course to esh Shejerah (alt. 795 feet), and Lubieh (alt. 920 feet). About half-a-mile north of esh Shejerah, the waterparting deflects to the south-east, and divides the basins of Wady Fejjas and Wady el Bireh. Judging from the observations on either side, as others are wanting, the altitude of the waterparting does not diminish between esh Shejerah and Sarona (alt. 892 feet). The range, however, proceeds from esh Shejerah to Lubieh, whence it follows a spur south-eastward to the cliffs on the north of Wady Mu'allakah. The wady intersects the range, which is taken up again at Kefr Sabt (alt. 650 feet), and runs on to Sarona. Two miles or more south-east of Sarona, the range passes el Hadetheh (alt. 735 feet), at the southern end of bold cliffs; and proceeds along the summit of the Kulah Cliffs (alt. 1,179 feet), which overhang the Fejjas gorge. The range now passes south and south-east to the junction of Wady el Bireh with the Ghôr of Jordan.

The Divisions of the Nazareth Range.

1. *The Western Division.*

Having traced the main range throughout, it may be examined more in detail under three divisions. The Western Division embraces the picturesque and quadrangular block of upland covered with oak forest, that lies between the Mukutt'a and the Melek. The boundary between this division and the next on the east, is to be found in the lowest depression or saddle that connects the Melek where it bends to the west, with the recess which the Plain of Esdraelon here makes northward towards the Melek and the Plain of

Buttauf. This depression would be better defined by one or two altitudes, which would be ample if taken in the right place or places.

The waterparting between the Mukutt'a and the Melek passes over this quadrangular block between its south-western and north-eastern corners. The Melek drainage has the longer slopes facing the north-east and north-west, and it forms a compact parallelogram, having the shorter slopes descending towards the Mukutt'a from its south-eastern and south-western sides. All the villages are on the south-eastern slope. The highest summit probably lies about half a mile north-west of Kuskus (alt. 575 feet), but its altitude is unknown.

2. *The Central Division.*

The central division of this range extends eastward to Wady el Mady, which descends to the southern base line on the east of Mount Tabor or Jebel et Tur. From the head of the perennial stream that washes Wady el Mady, the eastern limit of this division may be tracked northward along the Damascus Road, which runs nearly along the waterparting to the Ard ed Darûn, in the direction of the village of Lubieh and the Hakûl el Mughârah, a watercourse which descends to the Plain of Toron, and connects it with Wady Shubbâbeh, parts of the northern base line of this division. The ground around Lubieh does not fit well, either with the Toron or the Nazareth ranges, but on the whole it is found to fall in better with the latter, and the symmetry of the former is left undisturbed.

Between the northern and southern bases of the range, and the eastern and western limits of this division, the Nazareth Range presents its most complex features. The main range has been already traced. Looking at the one-inch map, it may be inferred that the main range runs to Semûnieh, whereas Semûnieh (alt. 623 feet), is situated at the end of a spur given off from the waterparting on the south of 'Ailût. There can be no doubt that more relative prominence is due to the hill shading of that waterparting range. On the south of Semûnieh, another prominent spur runs into the Plain of Esdraelon, and bears upon it the villages of Yafa and Mujeidil (alt. 780 feet), with the smaller hamlets of Jebâta (alt. 355 feet), and Ikhneifis at its base. The isolated mound of el Warakâny (alt. 277 feet), on the banks of the Mukutt'a, seems to be an outlier of this spur. The occurrence of this mound taken in connection with a similar elevation on the opposite bank of the river, accounts for the passage of a main road across this part of the plain, and also for the ancient prominence of Megiddo and Lejjun, as strongholds commanding the entrance into Samaria by this route.

Further east, the main range descends to the plain by gradually shortening slopes. About a mile and a half due south of Nazareth, the mountain descends to the plain by the rocky precipice of Jebel Kafsy (alt. 1,286 feet), which is represented by the Latin Church as the "Mountain of the Precipitation."* On the east of Iksâl, the edge of the plain is nearly within a mile of the summit of the range at Rujm el 'Ajamy. The head of the recess is about a mile and a half further east, near Debûrieh, and at this point, Mount Tabor (alt. 1,843 feet), is abruptly projected into the plain for more than two miles. A ridge from Mount Tabor towards the north-west, connects it with the main range at a point about half a mile north-east of the village of 'Ain Mahil, and the range soon makes an abrupt and short bend towards the north, and then another short bend to the north-east, where it throws off a spur towards the Wady el Mady, quite parallel to the ridge which connects Mount Tabor with the main range. The wadys which descend from the eastern side of the ridge, and from the main range between it and the spur, are all directed to a bed running along the western foot of the spur, which unites them in one outlet that joins Wady el Mady. The wooded flanks of Mount Tabor and the connecting ridge, skirt the Wady el Mady for about four miles

* St. Luke iv, 29. Guerin, "Galilee" I, 93 to 97. Rob." "Bib. Res." II, 335. Liévin, "Guide des Sanctuaires," 484.

and terminate the south-eastern part of this division, in striking contrast with the bare chalky downs spread out between Mount Tabor and the Sea of Galilee.

On its northern side, the main range sends out only one important spur westward from Jebel es Sih, (alt. 1,838 feet), on which stands el Meshhed (alt. 1,254 feet), and Seffûrieh (alt. 813 feet). This spur gradually spreads out like a fan, and throws off its lower ramifications from Rummâneh to el Khalladîyeh; its lateral limits being the Wadys Kefr Kenna and Rummâneh on the east, and a wady on the south, that has its sources in 'Ain el Jinnan and 'Ain el Jikleh, and descends by er Reineh, and Kustul Seffûrieh, which is an outwork on the south of the ancient town, the Sepphoris of Josephus. Thus the outlet of the Plain of Buttauf is gradually contracted to a narrow vale passing southward on the west of Seffûrieh.

Between Jebel es Sih (alt. 1,838 feet), and esh Shejerah (alt. 795 feet), the northern side of the main range descends to the Plain of Toran, with a slope that gradually abbreviates eastward from Wady Kefr Kenna to el Merhân, its breadth being three miles at the former, and only half-a-mile at the latter. On the north of esh Shejerah, the main range runs on to Lubieh, which has the northern base of the Nazareth range immediately below the village in the Hakûl el Mughârah, which descends to the head of the Plain of Toron, where the western slope on the south of Lubieh, also falls. The eastern slope between Lubieh and esh Shejerah, forms the head of a plateau between Mount Tabor, including its north-western ridge, and the eastern division of the main Nazareth range, the description of which follows.

3. *The Eastern Division.*

The eastern division of the Nazareth Range has its eastern base in the Ghôr of the Jordan, and the north-eastern in the Wady Fejjas and Wady Shubbâbeh. The western boundary meets the slope from the central division of the range between Lubieh and esh Shejerah and runs by Ard ed

Darûn southward to the Wady el Mády and the Wady esh Sherrâr. The last finally becomes Wady el Bireh, and forms the south-western base of this division.

An inspection of the features which dominate the right bank of the Wady Shubbâbeh and Wady Fejjas and bound the Plain of Ahma in that direction, will bring into view a long line of interrupted basaltic cliffs, which stands in advance of the Fejjas and el Bireh waterparting, as far as from Lubieh to Damieh and towards Sarona; while from Sarona towards the Jordan, the cliffs and the waterparting coincide. This is the eastern division of the Nazareth Range. It will also be observed that the base line formed by Wady Shubbâbeh and Wady Fejjas is remarkable for its depression below sea level. As far up as Damieh the depression is 84 feet, and no doubt it will be found to be still greater below that ruin, at the confluence of Wady el Mu'allakah with Wady Shubbâbeh. The depression is 300 feet at Kh. Seiyâdeh, near the junction of Wady Sarona with Wady Fejjas. It is about 640 feet below sea level at the junction of Wady Fejjas with the Jordan. The observation of a few points on the sea level around the Plain of Ahma, including especially Wady Shubbâbeh, would be an interesting contribution to the study of the Jordan Basin. At present the depression of this base line, and of the plain through which it passes, has only to be considered in relation to the summit of the eastern division of the Nazareth Range. The depression will be found to be in remarkable contrast to the comparatively elevated character of the Plateau of Sh'arâh, which extends from the summit of the range to Wady el Mády and Wady esh Sherrâr. The range is indeed a steep escarpment with rocky cliffs and precipices of basalt, repeatedly interrupting the smoother parts of the descent from the Plateau of Sh'arâh to the depressed Plain of Ahma. From Lubieh to Sarona, the range must be identified with a spur extending from Lubieh to the cliffs on the north of Wady Mu'allakah. That wady intersects the range between Kefr Sabt (alt. 650 feet) and Damieh. The descent of the wady from the plateau to Damieh is no doubt rapid, for the waterparting at the head of the wady is probably more than 700 feet above the sea, while at the foot of Damieh, the wady must be at least 100 feet below the sea.

At its southern end the range is extended from the cliffs of el Kulah (alt. 1,179 feet) due south towards the Wady el Bireh; but when it reaches about a mile from the wady, the range turns to the south-east and descends to the Ghôr, filling the angle between the Wady el Bireh and the Ghôr.

Besides the Wady el Mady, which flows along its western edge, the Plateau of Sh'arâh is watered by four perennial streams which rise near the line of cliffs. Three flowing southward to join Wady esh Sherrâr, are named Wady Sh'arâh, Wady Shomer, and Raud Tuffah. The fourth rises near Sirin and runs due west to join the Tuffah near its outlet.

Several villages are distributed over the plateau. Along the summit of the range are Kefr Sabt (alt. 650 feet), Sarona (alt. 892 feet), el Hadetheh (alt. 735 feet), and Sirin (alt. 570 feet). Towards the centre are Kefr Kama (alt. 650 feet), on a branch of Wady Sh'arâh; Madher (alt. 544 feet) between Wady Sh'arâh and Wady Shomer; and Meshah (alt. 320 feet) about a mile from the mouth of Wady el Mady.

From the heights of Mount Tabor and Jebel es Sih, which rise above 1,800 feet, the Plateau of Sh'arah is the first drop in the descent towards the Sea of Galilee, its mean altitude being about 600 feet. The second drop in the same direction, is the Plain of Ahma, which has been shown to be about 300 feet below the sea level in the centre of the plain. The surface of the Sea of Galilee itself is depressed 628 feet; and its greatest depth was reported by Lieutenant Lynch of the United States Navy to be 160 feet lower.

The higher hills on the west of the plateau, and the range which divides the Plain of Ahma from the Sea of Galilee, appear to be of limestone, which also seems to underlie the intermediate surface of the plateau and the plain. The line of interrupted cliffs between the plain and plateau is said to be basaltic, and lava currents with fragments of lava and pumice cover the whole region between Mount Tabor and the Galilean Sea. But the exactitude of the Survey makes it usually very difficult to apply the geological notes of former observers to it; as their representations of the ground seldom embrace the details which are now delineated, and their allusions are too general for precise use. Dr. Tristram's geological account of this part is remarkably distinct, but it would be rather presumptuous to attempt to colour the one-inch survey from that or any other existing description.* Professor Geikie's recent theory of the emission of lava from dykes as well as craters, could probably be studied under favourable circumstances around the Sea of Galilee.

The Jebel Duhy Range.

The southernmost group of the Hills of Lower Galilee is an irregular quadrilateral, which would be wholly confined to the basin of the Jordan, but for the north-western slope of Jebel Duhy or Little Hermon, which is within the basin of the Mukutt'a. The entire group has two faces towards the north, of which one is slightly inclined to the north-west, and has its base in the great arm of the Plain of Esdraelon, that stretches up between Jebel Duhy and Mount Tabor, as far as the end of the plain, and the head of the gorge of Wady el Bireh. The other northerly face is slightly inclined to the north-east, and descends in a deep valley to the Wady el Bireh. The third face is towards the Jordan and drops steeply to the Ghor of Jordan between Wady el Bireh and Nahr el Jâlûd. The fourth face inclines south-westward along the wide, sloping vale of Wady el Jâlûd or the Valley of Jezreel, between Beisân and Jebel Duhy. The length of the mountain on this line is about 15 miles. Its greatest breadth is about seven miles.

This block is distinctly divided into three parts. (1) The triangular or trihedral mass of Jebel Duhy occupies the western extremity and rises to the height of 1,690 feet. Its

* Tristram's " Land of Israel," 422. Lartet, " Exploration Géologique," 188.

northern slope contributes partly to the Mukutt'a, but chiefly to Wady el Bireh. The waterparting between the Mediterranean and Jordan watersheds, ascends to the summit of the mountain from its northern base, passing on the east of the village of Nein (where Our Lord restored the widow's son to life), and it descends westward by the main ridge to the Plain of Esdraelon and the village of el 'Afûleh. This village must be distinguished from its neighbour on the east called el Fuleh, as the latter is within the basin of Nahr Jâlûd, although both may have been within the compass of the towns that formerly stood here. Also on the northern side of the mountain are the miserable caves and huts of the village of Endor, where the wretched old women pour fanatical curses on passing Europeans, reminding them of Saul's visit there to the ancient sorceress.

Between Nain and Endor is a prominent hill with two summits named Tell el 'Ajjûl, formed of basaltic rocks, which at a distance look like ruins.* Dr. Tristram observed a basaltic or trap dyke at the south-western base of the mountain, and he attributes the elevation of the mountain to it, during the period when the basalt flowed over these and other parts of Galilee.†

East of Endor, a depression in the upland containing the village of Tumrah (alt. 680 feet) cuts off the ridge of Mount Duhy from the lower range on the east; and a spur sloping southward from Tumrah to Wady es Sidr, a branch of Nahr Jâlûd, combines with that wady to complete the eastern base of Jebel ed Duhy. The eastern slope is remarkable for an antique necropolis at Kh. Mâlûf.*

The south-western slope is drained by several branches into Wady el Hufiyir and Wady el Asmar, at the head of Nahr Jâlûd. Its base is spread out between Kh. Tub'aûn, on the north of Ain Jâlûd and el 'Afuleh on the main waterparting between the Mediterranean and Jordan. The latter place has been identified by Mons. Guerin with Aphek,‡ on

* Guerin, "Galilée" I, ch. vii, viii. † Tristram's "Land of Israel," 129.
‡ 1 Sam. xxviii, 4; xxix, 1; xxxi, 1.

the authority of the monk Buchardus. Aphek means a "stronghold" or "strength," and here has probably always been some indications of a fortified post. The Crusaders' castle of Faba is now in ruins at Fuleh. About the midst of this slope is Solam (alt. 445 feet) the Shunem of the Old Testament. All three places are historical. At Shunem* the Philistines collected their forces, while Saul pitched in Mount Gilboa. On the day of battle, the Philistines were drawn up on the plain at Aphek (el Afuleh) and the army of Saul took ground by a fountain in Jezreel, not necessarily Ain Jâlûd, and probably close to the Plain of Esdraelon. The distance between the two headquarters is four miles; and the battle would probably have been joined about midway or nearer Gilboa. On the defeat of the Israelites they fled for refuge to Mount Gilboa: and Saul with his sons were perhaps overtaken towards Beth Shan (Beisan); for on the walls of that city their bodies were exposed by the Philistines, and soon valiantly removed by the men of Jabesh, a neighbouring city across the Jordan.

From Shunem the beautiful Abishag was brought to nurse the aged David. It was also the scene of Elisha's repeated gratitude for the hospitality of a "great woman" who resided there.

In later times each end of this side of the mountain has been the scene of military events. At Tub'aûn (Tubania) was the Crusaders' camp in 1183, when Saladin was posted on the opposite side of the valley at 'Ain Jâlûd, and suddenly decamped. Between Fuleh and 'Afuleh on the 16th April, 1709, Kleber with about 1,500 men resisted an overwhelming Syrian force of 25,000. The French general maintained his ground during six hours from daylight to midday, when Bonaparte brought up a reinforcement of 600 men from Acre, and so manœuvred them that the enemy became alarmed, and Kleber assuming the offensive, stormed the village. The enemy then fled in great disorder, pursued by Murat, and

* 1 Sam. xxviii, 4; xxix, 1; xxxi, 1.

hundreds were drowned in a swollen stream, thus repeating an incident of Sisera's defeat.

A plateau on the east of Jebel Duhy embraces the remainder of this upland group. It is intersected almost diagonally by a winding wady named Wady Dabu and Wady Yebla after ruined sites on its banks. The wady rises on the northwest at Dabu near Tumrah, and it falls to the south-east where the wady emerges as Wady el 'Esh-sheh from the lofty banks of the plateau on to the Ghor. (2.) The northern part of the plateau is a slightly winding range, extending between Tumrah and the very prominent summit of Kaukab el Hawa, crowned by the vast remains of the Crusaders' great Castle of Belvoir. The general course of the range is parallel with the Wady el Bireh at its base, but while the wady descends to 900 feet *below* the sea level, the range rises gradually to 975 feet *above* the same datum line. The fall from the one to the other towards the east, or from the eastern wall of the fortress to the Ghor at its foot, is quite abrupt, the edge of the low plain being within the horizontal distance of a mile from the overhanging summit. The fall to Wady Bireh on the north, is broken by a terrace on which is the village of el Bireh (alt. 546 feet), the bottom of the great gorge being about 1,000 feet lower. The terrace widens out towards Denna and is terminated by a bend of the range towards the same place, from which a watercourse descends by Wady Hammûd to Wady Bireh. Between Wady Hammûd and the Plain of Esdraelon, the northern range of this group of hills descends to Wady Bireh, a broad and open down. Towards the south and south-west the fall varies. From Kaukab el Hawa there is a direct drop to the Wady el 'Esh-sheh, and its continuation westward as Wady Kharrâr. (3.) This valley forms the first step, terrace or raised trough, above the Ghor and the Valley of Jezreel. On the outer edge of the trough is Murussus (alt. 323 feet). Denudation has reduced the trough within this edge to an altitude of 10 feet only at Zebu. This, however, represents a material difference between the direct ascent to

Kaukab el Hawa on this side and on the side of the Ghòr. Further to the west the descent from the range takes place through the two parallel terraces or troughs, the uppermost being Wady Dabu, and the next lower being Wady Yebla and Wady el Hoktîyeh, below which succeeds the third parallel still lower, in Wady Kharrâr and 'Esh-sheh. The lowest of this parallel series is Nahr Jâlûd. Beyond the Wady el Hoktîyeh, that line of valley is prolonged westward, over a saddle, to branches of Wady es Sidr, leading up to Kh. Mâlûf.

As a conclusion to this introduction to the new Survey of Lower Galilee, a general view of its river basins, Lowlands, and Uplands, may be taken. Two great basins divide its Mediterranean watershed, the Nahr N'amein and the Mukutt'a. Six tributary basins drain the Jordan slope, namely, Wady Rubudiyeh, Wady el Hamâm, Wady Abu el 'Amîs, Wady Fejjas, Wady el Bireh, and Wady el 'Esh-sheh, besides Wady 'Amûd and the Nahr Jâlûd on the northern and southern borders. The lowlands of the Mediterranean slope are all embraced in the great plains of Acre and Esdraelon with the offsets of the latter including Buttauf and Toran. On the side of the Jordan, is the Ghuweir or Plain of Genessaret, the narrow strand of Tiberias, and the Ghor of the Jordan. The uplands are formed of four symmetrical ranges, of which the summits in particular are curvilinear and parallel. These are the Northern or Shaghûr, the Toran, the Nazareth, and the Jebel Duhy Ranges. The confronted slopes between these ranges, sometimes combine to form upland plains, like the Plain of Rameh, the Plain of Arrâbeh, the Plain of 'Ahma, the Plateau of Sh'arâh, and the Plateau of the 'Esh-sheh. Beautiful woodlands deck the scenery of the south-western slopes. Broad and open downs, bare and monotonous, even when rocky, meet the wearied eye on the south-east. In the north-east, Dr. Thomson describes the scenery between the Plain of Rameh and the head of Wady er Rubudiyeh as "exquisitely beautiful."

THE HIGHLANDS OF SAMARIA AND JUDÆA.

South of the line formed by the Rivers Mukutt'a and Jâlûd, no recognised features have hitherto served the purpose of dividing distinctly the long stretch of highland between Mount Carmel and Beersheba. Still on approaching the subject, it is only reasonable to expect that in the course of a hundred miles, there must be variations that admit of being conveniently grouped, and that should not be overlooked in a geographical description. But before the present survey, the best accounts of the country were too inadequate to enable any attempt of the kind to be carried out on the lines that will be now adopted. It was the great aim of Dr. Robinson's most able researches, " to collect materials for the preparation of a systematic work on the physical and historical geography of the Holy Land."* As much as it was possible for his genius to accomplish with the data at his command, has been fortunately preserved in the " Physical Geography of the Holy Land," made known in 1865 through Mr. Murray, by the tender hand on whom the duty devolved after the author's lamented death. This fragment will always be regarded by students with the attention and interest due to the last work of the most successful of all previous contributors to the geography of Palestine. Yet Dr. Robinson had to be content with little more than a general view of the subject, comprehending in one sweeping glance the whole region from Esdraelon to Hebron; and his details are confined to isolated accounts of the particular mountains mentioned in Scripture. The Survey of the Palestine Exploration Fund no longer allows the geographer to indulge in such a method. Every important feature is now exposed in its length, breadth, and height; and thus it has become quite practicable to discern certain natural groups and divisions that serve to bring to light the distinctive characteristics of the different parts of the country, and facilitate intelligible description and convenient reference.

* " Bib. Res." iii, Preface.

216   THE HIGHLANDS OF SAMARIA AND JUDÆA.

Five of such divisions have been adopted on the present occasion. The first has the Mukutt'a and Jâlûd Rivers on the north, and the Wadys Shair and El Humr on the south. The second division succeeds, and has on its southern boundary Wady Balût and Wady el 'Aujah. The third has for its southern boundary Wady Alallah, Wady Aly, Wady Ismaen, Sikreh and Werd, Wady el War and el Meshash. The fourth and fifth are divided between the Shephelah on the west, and the mountains of Judah on the east. The grounds of these divisions will be discussed in the following explanation of their details.

The Northern Samaritan Hills.

The north-eastern base of the Samaritan Hills, from the sea at the foot of Mount Carmel to the Jordan, is generally defined with precision by the abrupt termination of the slope in the continuous lowland formed by the plain of Esdraelon or Merj Ibn Amîr, the Valley of Jezreel or Wata el Jâlûd, and the Plain of Beisân. But for the considerable projection of Mount Gilboa to the north, the whole of this face would fall nearly in the same oblique line, running from north-west to south-east. Mount Carmel is slightly advanced beyond the central part of the range, and the foot of the hills on the south of the Plain of Beisân falls almost in the same line as the foot of Mount Carmel. Mount Gilboa projects between the Plains of Beisân and Esdraelon, along the line of the Mediterranean and Jordan waterparting, and thus brings prominently forward, the main axial division between the eastern and the western watersheds. It will be noticed that the western base of Mount Gilboa is nearly parallel with the western base of Mount Carmel; but the comparison leads to no inference. It may, however, be assumed that the recess between the two mountains is due to denudation of the softer chalky rocks which lie between those hard limestone masses, and which probably extended up to the line of the Mukutt'a River. Such an extension is still found running out to the river at Ludd, where the junction takes place of the head-waters of the Mukutt'a coming from the north-east and south-east. If the southern part of the Plain of Beisân has undergone a similar denudation, then the obliquity of the ancient base line would have been much less, and the protrusion of Mount Gilboa would not have existed.

The western face of these hills descends to the Plain of Sharon from Cape Carmel southward. The eastern face descends to the Ghôr of the Jordan. There are notable differences between the northern and southern parts of the Samaritan Hills, and the natural features enable a distinct dividing line to be drawn between them.

One extremity of this line is suggested by the conditions of the eastern slope. On the south of the Plain of Beisân, the hills advance to the banks of the Jordan, and soon exhibit a parallel structure in successive ridges and valleys, descending south-eastward to the Jordan from the main waterparting. The series terminates on the south in Wady el Ifjim, called Wady el Kerad towards its source on the edge of the Plain of Salim, and Wady el Humr in crossing the Ghôr. It should be remembered that the limits of mountains and hills, are to be found in plains or watercourses. Summits or ridges are not available for the purpose, as they cannot be naturally separated from their slopes.

On the north of Wady el Ifjim, the features of the ground are comparatively expanded and developed. On the south they are as remarkably contracted and reduced to small proportions. The northern part will be described at once. The southern part will follow in due course. It will then be seen that the Wady el Ifjim constitutes a dividing line between sections of country strongly contrasted with each other.

In the northern section, the summit of the Jordan slope is thrown back westward, as much as six to ten miles; and in a broad sweep it is also advanced eastward up to the Jordan in its central part; while it recedes from the river gradually towards the north and south. The slope is divided between four parallel ridges and their offsets. The most northerly ridge culminates in Ras el Bedd (alt. 1,760 feet); the next is capped by Ras Jadîr (alt. 2,326 feet); the third rises to Jebel Tammûn (alt. 1,960 feet); and the fourth predominates in Jebel el Kebîr (alt. 2,610 feet). To these must be added the range between the Maleh Basin and the Jordan, which rises in Ras Umm Zokah to 840 feet above the sea, or 1,930 feet above the Jordan.

The interior valleys of this part of the northern section are included in the basins of Wady el.Maleh, Wady el Bukei'a, and Wady Fâr'ah. The exterior are found on the slopes towards the Plain of Beisân, the Jordan, and the Wady el Ifjim.

The prolongation westward of the line of the Ifjim and Kerad, is distinctly found at the base of the same ridge, in the mid-channel of the Plain of Salim; and further west, in the watercourse which joins it through Wady esh Shejûr, from the great gap between the mountains of Ebal and Gerizim, in which the city of Nablûs is situated. Westward of Nablûs, the line is taken up and carried to the Plain of Sharon by the Wady esh Shair, or Zeimer, which lies at the southern base of a mountain range on the north, that culminates in a summit dedicated to Sheikh Beiazid (alt. 2,375 feet). The Wady esh Shair is continued to the sea as Nahr Iskanderuneh, and throughout defines the limit of features as distinct of their kind as those on the eastern slope. To trace these in order, it is necessary to recur to the description of the eastern slope, and connect it with the northern and central parts of the section.

The northernmost of the ranges on the eastern slope, culminating in Ras el Bedd, joins the main waterparting at Ras Ibzik (alt. 2,404 feet). Its slope to the Plain of Beisân is a portion of the northern escarpment of the Samaritan Plateau. From Ras Ibzik, the summit of the scarp is continued along the main waterparting to Tannin (alt. 1,460 feet). From Tannin to Mount Carmel, the same line of summit is found along the south-western waterparting of the Mukutt'a Basin.

MOUNT GILBOA.   219

Mount Gilboa or Jebel Fukû'a.

From Tannin and Ras Ibzik respectively, two ranges advance from the Samaritan Plateau in northerly and convergent directions; and they unite at Jebel Abu Madwar (alt. 1,648 feet). These embrace the southern part of Mount Gilboa, and form the separate summits of its south-eastern and south-western slopes. The village of Jelkamûs (alt. 1,308 feet) is on the western range. From Jebel Abu Madwar the mountain is prolonged as a single range northward to Sheikh Barkan (alt. 1,698 feet). Then it takes a west-north-west course with a slight curve to its western end, where it divides into three spurs, which descend to the plain at Zerin on the north, at Sundela on the south, and at an intermediate point on the road between those places. Between Sheik Burkan and Jebel Abu Madwar, a broad and undulating branch is thrown off, which extends somewhat further west than the main range. It is occupied by the villages of Mukeibileh (alt. 313 feet), Jelameh (alt. 352 feet), Arrâneh, Deir Ghuzaleh (alt. 738 feet), and Fukû'a (alt. 1,502 feet). The last is the place from whence the local name of the mountain (Jebel Fukû'a) is now derived. A spur of less length descends from Jebel Abu Madwar to Beit Kâd (alt. 687 feet).

The slope of Mount Gilboa towards the Valley of Jezreel and the Plain of Beisan, is much shorter than its western side, it is indeed quite abrupt, and with little variation, except at Wady el Judîd, and Nuris, near Zerin.

Between the two ranges of Mount Gilboa, which diverge southwards from Jebel Abu Madwar, there is a distinct series of elevated valleys, sometimes on one side of the main waterparting, sometimes on the other, and occasionally on both sides. Such valleys running laterally to the main waterparting range, are not confined to this part, for they will be found with unlimited variations of form and development, and with scarcely any interruption, as far southward as the Survey extends. Their importance in many cases, in facilitating communication between north and south cannot be overrated.

The first in the series is the Valley of Jelbon (alt. 1,062 feet) which measured between the summit of Jebel Abu Madwar and the saddle which terminates it on the south, is three miles in length. It lies on the western side of the main waterparting, and its drainage passes off through a gorge in the western range, and reaches the Plain of Esdraelon as Wady en Nusf. This valley is divided from the next on the south by the transposition of the main waterparting to the western side of the next valley. The elevation of the valleys makes the separation slight, and their connection is maintained by a road across the low saddle that divides their waters.

The second valley is of greater extent, more curious in its structure, and divided into two distinct parts with a village in each. The main waterparting, in crossing from east to west, forms its northern boundary, and divides it from the Vale of Jelbon (Gilboa). Continuing to the south-west as far as Tannin, the parting bounds the valley in that direction, and has the village of Jelkamûs (alt. 1,308 feet) on its summit. At Tannin, the main waterparting bends sharply to the south-east, as far as Ras Ibzik, where it is again on the same line that it described on the east of Jelbon. The Jelbon range, though it ceases to be the waterparting, is not lost, for it passes straightway to Ras Ibzik, and forms the eastern enclosure of the elevated valley under notice, as well as the summit of the slope that descends from it to the Plain of Beisân.

The interior of the second valley forms two divisions so distinct that they must bear separate names derived from their respective villages, El Mughair on the north, and Raba on the south. The drainage of the northern part originates on the eastern range, and runs westward along the northern waterparting, till that feature bends to the south-west, and causes the drainage to take the same direction, passing the village of El Mughair (alt. 1,072 feet). The drainage of the southern part also proceeds from the eastern range, near the village of Raba (alt. 1,590 feet), and runs in two parallel channels also westerly, and along the foot of the waterparting range, where it extends between Ras Ibzik and Tánnin. The western boundary which turned the Mughair Wady to the south-west, diverts the Raba Wady after the junction of its parallel streams, to the north-east, that is in the same line as the Mughair Wady, though they proceed from opposite directions. The streams are thus brought to a junction, and acquire the name of Wady Shubash, which intersecting the eastern range in a deep gorge, descends rapidly to the Plain of Beisân on an easterly course.

The next elevated lateral valley on the south of Raba, and east of the main waterparting, extends from Ras Ibzik to Tubas (alt. 1,227 feet), and is drained by streams running in the same line, but from opposite directions, at the western extremity of the Maleh Basin. From Tubas, the succession is maintained by the wide-spreading heads of Wady Farah, which repeats the eccentric features of the Mughair-Raba Valley, on a scale three times greater, and extends the lateral communication along the eastern side of the main waterparting from Tubas to the Plain of Mukhnah, a distance of 12 miles. From the last named point, the series of these elevated plateaus will be taken up again in due course.

If nothing else served to fix the southern boundary of this section towards the west, the drainage proceeding from the northern edge of the plateau in this direction, would answer the purpose. It belongs to the basin of the Nahr el Mefjir, which has its southern waterparting along the Beiazid Range. As the southern base of the Beiazid Range, the Wady Shair becomes the boundary of this part of the section, and thus its continuity with the eastern part of the base-line, is shown to be not so much a motive as a coincidence.

When Wady Shair or Zeimer reaches the foot of the hills, the base line of the highlands is continued northward along the Plain of Sharon to Cape Carmel. But regarding the low eminences in the plain as related to the highland, it becomes necessary to prolong the boundary of the section across the plain along Wady Shair, to its junction with the Nahr Iskanderuneh, and to the outfall of the latter into the sea. The coast line then forms the boundary of this enlargement.

The whole of the Western Plateau of the Northern Samaritan Hills from the Beiazid Range to Cape Carmel, may be thus divided according to its natural features :—

(1). Mount Carmel, having its southern base in Wady el Mihl and Wady el Matabin, called lower down Wady Henu. This mountain with its parallel ranges has been so fully described in connection with its watercourses, as to require no further notice. See pages 26 to 30.

(2). The lower tract of Belad er Ruhah, with the corresponding slope to the Mukutt'a. This may be conveniently limited on the south by Wady 'Arah, the north-western branch of Nahr el Mefjir, descending westward from Umm el Fahm, and eastward by the wady which skirts Musmus, and the high road in that direction. This tract descends on the west, partly to the narrow Plain of Tanturah, and partly to the broader Plain of Sharon. The descent is abrupt to the narrow plain, and forms the bold headland of el Kashm. It is chiefly in the basins of Nahr ed Dufleh and Nahr ez Zerka. See pages 30, 31.

(3). The wooded heights culminating in Sheikh Iskander (alt. 1,690 feet), embracing the rugged hills of Umm el Khataf, and extending southward to the Wady of Burkin, the Plain of 'Arrâbeh, and Wady el 'Asl. These hills are the Northern Heights of the Mefjir Basin. See page 34.

(4). The Eastern Plains of the Mefjir Basin, including the Plain of 'Arrâbeh or Dothan, the plateau of the Upper Selhab; the Merj el Ghurûk, and the Plain of Fendakumieh.

(5). The Western Heights of the Mefjir Basin, culminating in Batn en Nury (alt. 1,660 feet).

(6). The Southern Heights or the Beiazid Range and Jebel Eslamiyeh, the Mount Ebal of Scripture.

For a description of these six divisions in more detail, reference may be made to the account of the basins which

THE NORTHERN SAMARITAN HILLS. 223

contain them. See pages 30 to 34. But it may be well, in addition, to point to the connection of the Plain of Esdraelon, the Eastern Plains of the Mefjir Basin, and the uplands of the Jordan Slope from the Plain of Salim to the Plain of Beisân. These form an unbroken succession of pastoral plains and downs.

The following routes across the northern part of the Samaritan Hills also claim attention. Lateral valleys along the summit of the Jordan slope are traversed by the ancient highway between Nablûs, Beisân, and the Sea of Galilee. The road from Nablûs to Jenin and Nazareth, skirts the Plain of Fendakumieh, and the Merj el Ghurûk, and passing through Kubatieh, it has the plateau of the Upper Selhab on the right, and the Plain of 'Arrâbeh or Dothan on the left. From the north-eastern edge of the Plain of 'Arrâbeh, there is but a short descent of less than a mile, to the Plain of Esdraelon, by way of Burkin. The descent from the Plain of 'Arrâbeh to the Plain of Sharon is about nine miles in length, and the road takes a hilly track by Ferâsîn (alt. 787 feet) and Kuffin (alt. 460 feet). The upland between the Plains of Esdraelon and Sharon is crossed by a highway which leaves Kerkur on the west, and ascends the Wady 'Arah, for about four miles from the foot of the hills, when it forks, sending two branches to the Plain of Esdraelon, through Musmus to Lejjun on the right, and through Kefrein on the left. The high road from Kerkur to Acre, proceeds at first due north, over wooded hills to Kannir (alt. 275 feet), and near Umm esh Shûf (alt. 400 feet) ; thence it advances more easterly across the open chalky downs of the Belad er Ruhah, and descends to the Plain of Esdraelon, at the mouth of Wady el Mihl, and the foot of the isolated mound of Tell Kaimoun, once fortified. The nearest habitation is the Druse hamlet of el Mansurah, at the foot of Mount Carmel ; the whole length of this route across the hills is about 14 miles, between Umm esh Shûf and Mansurah. The only village is Daliet er Ruhah (alt. 729 feet), about half a mile on the east of the road.

The Southern Samaritan Hills.

The southern limit of the Samaritan Hills, and consequently of this section of them, will be traced from the sea along the Nahr el 'Auja, and the Wady Deir Balût, which emerges from the hills at Mejdel Yaba.* At Kurawa ibn-Zeid (alt. 2,130 feet), the Wady receives two chief affluents from the north-east and south-east respectively. The north-eastern wady takes several names, including Wady Ishar. The south-eastern branch, named Wady el Jib, is chosen for the present purpose. It meanders in a deep and sinuous channel as Wady el Jib and Wady en Nimr, and skirts the foot of the steep and lofty Sinjil Ridge, well known to travellers between Jerusalem and Nablûs. The Wady en Nimr rises on the Mediterranean waterparting at the northern base of Tell 'Asur (alt. 3,318 feet), and descends between the villages of Mezrah esh Sherkiyeh on the north, and Selwâd on the south, to join the Wady el Jib, at the bottom of the Sinjil ridge. The Wady el Jib rises on the top of the plateau, near the village of Sinjil (alt. 2,600 feet), and makes a gradual descent to the valley by a gorge through the ridge, the slope being here three miles in length, with a fall of 600 feet. Westward the ridge stands out abruptly; and about two miles and a half from Sinjil, at the village of Jiljilia (alt. 2,441 feet), the drop into the valley is 700 feet, within the horizontal distance of one mile.

From the source of Wady en Nimr, the main waterparting is crossed to continue the line along a brook which has its origin at three-quarters of a mile south-east of Mezrah esh Sherkîyeh, and descends, first to the north-east and then to the south-east, to fall into the rocky gorge of Wady Samieh, and thus reach the enclosed plain, which has been supposed to be identical with the Vale or Plain of Keziz.† The Wady receives a permanent stream from 'Ain el 'Aujah, and with the name of Wady el 'Aujah it runs across the Plain of Keziz, the

* See Conder's "Tent Work," ii, 227.
† See pages 81, 82.

low hills, and the Ghor, to join the Jordan, where that river is 1,200 feet below the sea.

It has been already stated that the eastern extremity of this section, displays a remarkable contrast in its natural features, to the corresponding part on the north. The ranges of twelve miles in length, with the fine open valleys on the north, here give way to mere spurs and ravines, corrugating the face of an abrupt slope, of three or four miles in length, and rising within that distance, to about 2,500 feet above the depressed Ghor. This abrupt transition is explained by another contrast arising from the sudden projection of the main waterparting towards the east, for as much as six miles, a phenomenon which takes place between Jebel et Tor (Mount Gerizim), and the summit of et Tawanik (alt. 2,847 feet).

The eastward advance of the waterparting, terminates above the Wady Samieh, where the southern limit of this section has been drawn. The recession of the waterparting to the westward, is marked by the occurrence of a curious feature, which has been brought to light by the survey. It is the small basin with no outfall, named Merj Sia, about a mile and a half in length and breadth, and surrounded by eminences rising from 2,510 feet to 2,835 feet. Besides the change in the direction of the waterparting, other features also serve to distinguish this eastern side of the southern part of the Samaritan Hills, and the highland that follows it on the south. The distinction in that direction, will be found in the account which will follow in its place, no less striking than the contrast already displayed between it and the northern part.

Amid the great contraction of the width of the Jordan Slope in this section, the elevated lateral valleys to which attention has been directed, from Mount Gilboa southward continue to be found. Next to the high Plain of Salim, a valley commences at the southern foot of et Tawanik (alt. 2,847 feet) and continues to skirt the main waterparting as far south as Mejdel-beni-Fadl (alt. 2,146 feet). Its length exceeds five miles. It formed, no doubt, a part of the Roman Toparchy of Acrabatene, for the modern village of 'Akrabeh (alt. 2,045 feet), with antique remains, is found upon the main waterparting about midway between the ends of the valley which it completely dominates. The village of Yanûn in the northern part of the valley, is considered to occupy the site of the biblical Janoah. The village of Mejdel-beni-Fadl (alt. 2,145 feet) overlooks the southern part.

The upland valley appears to be called the Jehir 'Akrabeh and to be divided into an upper and lower terrace or trough parallel to each other. The upper includes three different drainages, descending severally from et Tawanik, 'Akrabeh, and Mejdel, the first passes by Yanûn and the Wady el Abeid to a stream of the lower tier, which also comes from et Tawanik; and the others fall to a distinct lower branch, called Wady es Seba. These form a final junction about midway between the extremities, and then bend suddenly to the east to intersect the outer range, and descend by Wady ed Dowa and Zamur, to the Plain of Ifjim.

Next to the Jehir 'Akrabeh is the Wady Nasir, on the west of Mejdel. Then follows a more extensive system, which has its northern extremity on the west of the village of Domeh (alt. 2,006 feet); the southern being defined by a ridge occupied by the village of Mugheir (alt. 2,246 feet). About a dozen different wadys contribute to two main drains called Wady ed Duba and Wady el Merâjem, which descend from the opposite ends of the valley to meet midway, and then intersect the outer range, and drop down some 2,000 feet to the Ghor, at the northern end of the ruins of Phasaëlis (Fusail). It is the outer range which forms the culminating summit of the eastern slope. Mugheir is on the waterparting between this valley of many names and Wady Samieh. The parting runs back to the westward for about four miles, and separates the eastern parts of the Highland of Samaria and Judæa. Dr. Robinson traversed the whole series of the upland valleys from Mugheir to the Plain of Salim, in the contrary direction to that which has been now followed. He gives an excellent account of them.* Mons. Guerin visited the same tract, and treats with more minuteness than any other authority on the antiquities, the villages, and their inhabitants, including some points not to be found in the survey.†

On the west of the main waterparting, this section is throughout an undulating plateau, the western edge of which, as observed from the Plain of Sharon, would probably be represented by summits of 1,200 feet, or thereabout, with variations descending to 1,000 and rising to 1,500. Towards the main waterparting, the altitudes frequently exceed 2,000 feet, and culminate in Kh. el Kerek (alt. 3,002 feet). It may be well to note the small Plain of Mukhnah, because it lies at the eastern base of Mount Gerizim, and is so well known to travellers between Jerusalem and Nablûs; but it is insignificant in comparison with the larger plains on the north. There is also the straight Wady Ishar, running south-westward from 'Akrabeh to Kurawa, with scarcely a bend for 13 miles. It receives no affluent of any consequence from its right bank, but on its left bank it takes all the drainage between it, the Sinjil Ridge, and the main waterparting.‡

The range between 'Akrabeh and Iskaka, throws off its waters northward to the Wady Kanah, and sends a spur between that wady and Wady Yasuf, as far as their junction. Another spur proceeds from Iskaka to the Kanah, and has the villages of Merda, Kefr Harîs, and Deir Estea on its summit. Two ridges emanating from Kefr Harîs, run to the Plain of Sharon, and enclose the Wady Rabah and its branches, dividing it from Wady Kanah, on the north, and Wady Ballût on the south.

* Robinson's "Bib. Res." iii, 292-297.
† Guerin, "Samarie," ii, ch. xxix-xxxii.
‡ Probably the Wady Ishar once formed a part of the political boundary between Judæa and Samaria. Although it seems to the present writer that there can be no question about the southern limits of the natural section, which is included herein, under the name of the Samaritan Hills; yet it should be observed that only a qualified political significance is to be attached to it. It is not intended here to enter upon the question of the political limits of Samaria in former times.

Like Wady Ishar, the Kanah generally hugs its right bank, which is formed by a range extending from Jebel et Tor or Mount Gerizim, to the Plain of Sharon at Hableh. This range affords a gentle incline between the Plain of Sharon and Nablûs, and is consequently followed by a highway. Between Jebel et Tor and Funduk, the range throws off its waters northward to the Wady et Tin. But from Funduk to Hableh, its northern drainage goes to Wady Kalkilieh. A spur from Funduk divides Wady Kalkilieh from affluents of Wady et Tin, and also from the Wady en Naml, which belongs to the Kalunsaweh branch of the Iskanderuneh Basin. Indeed this spur and the range between Funduk and Jebel et Tor, form together the waterparting between the Iskanderuneh and el 'Auja Basins.

In the Plain of Sharon this section includes an isolated group of low hills, bounded on the north by the Nahr Iskanderuneh, and on the south by Nahr el 'Auja. The Kalunsaweh branch of the former, and the Kalkilieh branch of the latter divide the group from the foot of the highland. It has been already noticed, as a part of the Plain of Sharon, with which it is usually associated.* But it seems probable that its connection with the highland was formerly closer; and considering a future examination with reference to that relationship to be desirable, attention is here invited to it. These Falik hills support a considerable population, of bad character, but rich in horses, flocks, and herds. Mukhalid, or Umm Khalid, was a station of the Survey, with a fine view from Mount Hermon to Jaffa. Lieutenant Conder mentions an open woodland as still existing around Umm Sur, and both he and Mons. Guerin call attention to it as a probable remnant of Assur Forest, where King Richard I overcame Saladin in a great battle on September 7, 1191, under the walls of Arsuf, the ancient Apollonia.†

* See *ante*, p. 15.
† " Tent Work " ii, 213, 219 ; Guerin, " Samarie " ii, 374–388.

THE MOUNTAINS OF JUDÆA.
Northern Group.

From the Jordan to the Plain of Sharon, the southern banks of Wady el 'Aujah, Wady Samieh, Wady el Jib, and Wady Deir Balût,—define the northern limit of the mountains of Judæa, including a part of Mount Ephraim, as well as the heights of Judah and Benjamin. Perhaps the southern edge of Mount Ephraim reached to Tell 'Asûr, Yebrûd, Bir ez Zeit, and Batn Harasheh.

The summit of the slope descending to the northern limit of the Judæan Mountains, is distinguished by the following culminating points from east to west :—(1.) En Nejmeh (alt. 2,391 feet), on a spur eastward from Tell 'Asûr, and dominating the Plain of Keziz and Wady el 'Aujah ; (2.) Tell 'Asûr (alt. 3,318 feet), on the Mediterranean waterparting ; (3.) A summit south-west of 'Attâra (alt. 2,791 feet), which belongs to a ridge continued through Umm Suffah (alt. 1,997 feet), Deir en Nidhan (alt. 1,934 feet), and 'Abûd (alt. 1,240 feet), to the Plain of Sharon.

The height of 'Attâra is also connected with a ridge running southward to the summit occupied by the ruins of Kh. Bir ez Zeit (alt. 2,665 feet), which is the centre of a system of ranges that enclose the two branches of Wady esh Shellal (Surar or Budrus), that unite at Shebtin. These heights divide the two Shebtin valleys and their branches : (1.) On the north, from Wady el Jib and the heads of the Wadys that meet at Deir Nabala, near the Plain of Sharon ; (2.) On the south, from the northern branch of Wady Malâkeh,* which joins the wady from Shebtin, on the west of Nalin ; (3.) On the east, from affluents of Wady el Jib, which descend to it northwards from Beitin and Abu Kush, to join the Jib on the east of 'Attâra. These affluents of Wady el Jib, form a noted section of enlarged systems of lateral communication along the Mediterranean waterparting, which extend throughout the remainder of the survey southward.

* Called also Shamo, Dilbeh, and Hamis.

The high road between Jerusalem and Nablûs, traverses them. They separate the Mediterranean and Jordan waterparting between Tell 'Asûr and Beitin, from the equally high and parallel heights of Bir ez Zeit.

The Wady Malâkeh, from its source at Ramallah, to its junction with Wady Shellal at Medieh (the Maccabean Modin), and afterwards the latter wady, and Wady Nusrah, may be regarded as dividing the hills on the north from a region on the south with distinct features. The corresponding part of the Jordan slope is bounded by the Wady Muheisin, which rises near Beitin, and is continued by Wady Rummaneh to Wady Nûei'ameh.

Westward of the main waterparting, the hills and valleys on the north are very intricate and convoluted. Besides those surrounding the two Shebtin valleys, which also form the western boundary of the lateral system running northwards from Beitin and Abu Kush, attention to the following parallel features may help to throw light on the mass. The western slope ascending from the Plain of Sharon, appears to culminate, firstly, in a range of summits represented by 'Abûd (alt. 1,240 feet), Deir Abu Meshal (alt. 1,556 feet), Shubkah (alt. 1,058 feet), and Deir el Kuddis (alt. 1,264 feet) ; (2.) Eastward, at a distance of three or four miles, rises a parallel series of superior heights, represented by Kefr 'Ain (alt. 2,285 feet), Kh. Kefr Tat (alt. 2,770 feet), Neby Saleh (alt. 1,868 feet), Deir en Nidhan (alt. 1,934 feet), Beit Ello (alt. 1,797 feet), Deir Ammar (alt. 1,737 feet), and Ras Kerker (alt. 1,637 feet). The interval between this range and the western series may be regarded as an intermediate terrace ; (3.) About five miles further east is the Bir ez Zeit range ; (4.) Then follows after an interval of three or four miles, the main waterparting from Tell 'Asûr to Beitin ; (5.) Finally, the Jordan slope exhibits the easternmost of these parallel ranges in the culminating summits of the ascent from the Jordan Plains, which are represented by en Nejmeh (alt. 2,391 feet), in continuation of a ridge running southward from el Mugheir, and by Khubbet Rummâmaneh (alt. 2,024 feet).

Between these easternmost heights and the main range, is the succession of lateral valleys beginning on the south of el Mugheir, with those which meet in Wady Samieh; followed by the heads of Wady Dar Jerîr; and the remarkable group descending southwards from Tell 'Asûr to Wady Mueheisin and Wady Asis, at the head of Wady Nûei'ameh. The interval between these ranges is about five miles in width. It is doubtless a part of the Wilderness of Beth Aven. The main wadys cut through the outer range in deep and rocky gorges, which become exceedingly steep on the final descent.

On the west of this northernmost part of the Judæan highland, must be included the isolated group of low hills in the midst of the Plain of Sharon, between the Nusrah branch and the main stream of Nahr el 'Auja. They reach an altitude of 295 feet.

The Jerusalem Group.

South of the foregoing division, the Judæan heights are characterised very remarkably and distinctly. Beginning on the west, it will be seen that a marked separation occurs on the south of Wady Malâkeh, between the base of the mountains rising toward the east and the lowland hills sinking toward the west. On the north of Wady Malâkeh, the descent from the main waterparting to the Plain of Sharon is gradual, and the successive terraces already indicated, succeed each other with comparatively little interruption of the general slope. But on the south of Wady Malâkeh, "the frowning mountains of Judah rise abruptly from the tract of hills at their foot;"* and the highland is divided from the lowland hills by a succession of valleys running north and south, or in a meridional direction. The most northerly of this series is Wady el Muslib, a branch of Wady Malâkeh. It is followed by Wady el Mikteleh, which runs south through the beautiful Plain or Merj of Ibn Amir, Joshua's "Valley of Ajalon."† This plain is more than two miles

* Robinson ii, 231.
† Josh. x. 12.

wide; it strongly emphasises the division between the highland and lowland, and is one of the characteristic features of this part. The continuity of the meridional valleys is broken at Yalo, but it is soon renewed again by a valley descending from the south towards Latron, and by another running southward to Wady es Surar, between Surah (Samson's Zorah), and Artuf. At Artuf, the Wady en Najil falling into Wady es Surar prolongs the meridional series, which is carried still farther south by affluents of Wady es Sunt, especially Wady es Sur. The distance from Wady Malâkeh to the head of Wady es Sur is about 24 miles.

The distinction between the lowland hills and the mountains on the east of them, is definitely expressed by the series of altitudes observed along the high road between Jaffa and Jerusalem. The plain at the foot of the hills near Ludd, is 165 feet above the sea; and the rise goes on steadily for eight miles up to 940 feet at Bir Main, or 960 feet, a little beyond. Here the hills drop down suddenly to the Plain of Yalo (Ibn Amîr or Ajalon), where the lowest observation is 685 feet. East of the plain, the mountains of Judah rise rapidly, and reach an altitude of 2,172 feet, near Beit Anan, about two miles from their base on the edge of the plain, or four miles from the level of 685 feet before mentioned. About three miles further east, the altitude is 2,621 feet at Beit Izza. This place is on the waterparting which divides the basins of Nahr el 'Auja and Nahr Rubin. Here this waterparting also coincides with the western boundary of the great Plain of Gibeon, or el Jib, drained by the Wady Beit Hannina and its affluents. The eastern edge of the plain is the main waterparting between the basins of the Mediterranean and the Jordan. The Plain of el Jib is thus on the summit of the Highlands of Judæa; and the direction of its drainage designates it as a portion of the series of great lateral valleys, which, while they form parts of the general slopes on either side of the main axis—lie at right angles to the slope, and parallel with the axis, and thus greatly facilitate communication.

The northern limit of the Plain of el Jib in its fullest

extent, is the waterparting of the Rubin Basin, from Bireh to Ram Allah; or it may be confined to the foot of the slope descending southward from that eminence. Its southern edge has been already described in page 43, on the authority of Dr. Robinson. But it seems preferable for the present purpose to take a larger view of the subject, and to connect with the Plain of el Jib, the ground on the west of Jerusalem, and also the Plain of Rephaim, called on the survey, el Bukei'a, which lies on the south-west of the city. The whole may come under the general denomination of the Western Plateau of Jerusalem. There is an Eastern Plateau adjoining on the Jordan watershed.

The group of the Judæan Hills now under notice, is limited on the south by another group, named the Shephelah in the Bible, and corresponding in extent with ancient Philistia. The low hills of the present group are of a similar character to the Shephelah, and adjoin that tract; but they are clearly related to the Plain of Sharon, while the hills of the Shephelah terminate in the range including Surar and Abu Shusheh, and dividing Sharon from Philistia. The towns of the Shephelah and Philistia named in the Bible, so far as they have been discovered, all lie to the south of the Plain of Sharon, which does not extend southward beyond the basin of Nahr el 'Auja, including the valleys which pass to the plain by Ramleh and Ludd. The dividing line between this mountain group and the Shephelah, is therefore traced along the watercourses which pass south-eastward from Ramleh by el Kubab and south of Latron, to the waterparting on the north of Eshua, and then by the watercourse which passes Eshua, to Wady es Surar. Here the borders of the Shephelah are left, and Wady es Surar is ascended as far as its confluence with Wady es Sikreh, when the latter is followed upwards to the junction of Wady Ahmed, which is next ascended up to the Jordan waterparting on the west of Bethlehem. On the east of the waterparting, the dividing line is laid along the wady named et Tahuneh, Fureidis, Khureitun, el Mu'allak, and ed Derajeh, to the Dead Sea.

The line from Wady es Surar to Wady ed Derajeh, is chosen for the purpose of dividing the mountain group distinguished by the Plain of Yalo and the higher plains of the Plateau of Jerusalem, from the more elevated and otherwise distinctly characterised group, which has its centre about Hulhul and Hebron, and is called in the Bible "The Hill Country." South of the line, the ground rises at once. At el Khudr, or the Convent of St. George, it is 2,832 feet, and at Tekua it is 2,798 feet, with a continual ascent to 3,546 feet at er Rameh on the north of Hebron. On the north of the line, the prominent Frank Mountain, opposite Tekua, is only 2,489 feet; and the Convent of Mar Elyas facing el Khudr is 2,616 feet. No heights of 3,000 feet are found northward until Tell 'Asûr is reached.

The foregoing remarks on the present group, have been directed mainly to the purpose of expounding the main features that give occasion for the group, and the determination of its limits. It is now proposed to regard the Western Jerusalem Plateau, including the Plains of el Jib and Rephaim, as the centre and summit of the group, and to take note of the slopes which descend therefrom to the west (1), south (2), and east (3).

(1.) It has been already observed that the waterparting between Nahr el 'Auja and Nahr Rubin, forms the western edge of the Plain of el Jib. The drainage from the northern and western side,—between Bireh, Ram Allah, and the Beth Horon Road—is carried off to Wady Malâkeh by Wady Hamis, Wady el Kelb, Wady 'Ain 'Arîk, Wady es Sunt, and Wady el Mahbis and Imeish.*

Mountain spurs, interposed between these wadys, descend from the Plain of el Jib, to the left bank of the Malâkeh, and have their heads at Ram Allah (alt. 2,850 feet), el Muntar, (alt. 2,685 feet), and Beitunia (alt. 2,570 feet). The Wady Imeish has its head in the Plain of el Jib, and in its descent, it becomes the deep gorge on the north side of the famous pass dominated by the villages of Beit Ur el Foka,

* See *ante*, page 39.

and el Tahta, or Beth Horon, upper and lower. The pass is along the ridge of a spur, which has an altitude of 2,545 feet above the sea on the edge of the Plain of el Jib. It divides into three at its base below Beit Ur et Tahta. The Jerusalem Jaffa road, comes to the ridge from el Jib, and follows it to Lower Beth-Horon, where it continues its descent by the southernmost prong of the triple fork to Beit Sira in the Plain of Yalo or Ajalon. No less than three main roads, and six minor roads meet at Beit Ur et Tahta.

On the south of the Beth Horon ridge, is Wady Selmân, Suleiman, or Solomon, which also rises on the western edge of the Plain of el Jib, and descends to the Plain of Ajalon, on its way to el Kubab, Ludd, and Nahr el 'Auja. Another high road between Jerusalem and Jaffa, runs along the bottom of this gorge to the Plain of Ajalon, where it crosses one of the Beth Horon roads, and runs on to Berfilia, Ludd, and Jaffa. About a mile west of Jimzu, this road falls in with the parallel road on the north, from Beth Horon to Ludd, which passes under the ruins of Medieh, identified with Modin of the Maccabees. A spur commencing at the village of Beit Izza (alt. 2,620 feet), on the west of el Jib, forms the southern side of the upper part of the Selmân gorge, and divides it from Wady el Marud or Keikabeh, an affluent which rises on the edge of the Plain of el Jib, between Beit Izza, Biddu, and el Kubeibeh, and unites with the Selmân at the foot of Beit Ur el Foka. Below the junction of Wady el Marud, Wady Selmân has for its southern bank, a spur that descends north-westward from the waterparting at Biddu, between Wady el Marud and Khallet el Kala, and originates in the prominent cone of Neby Samwil (alt. 2,935 feet), which rises in the midst of the Plain of el Jib. Another road from the Plain of Yalo to Jerusalem ascends this spur and passes Beit Likia, Beit Anan, el Kubeibeh, and Biddu.

From Kubeibeh, a range runs westward to the Plain of Yalo, with a broad slope to the north, but without any place of note; another spur from Biddu, passes due west to the Wady el Kotneh and the village of Katanneh. At its extremity is the ruin of Kefireh, the Chephirah of Joshua ix, 17.

South of Biddu, the waterparting of the 'Auja and Rubin Basins ceases to be connected with the Plain of el Jib, although it continues to throw spurs from its northern side, westward towards the Plain of Yalo, and to 'Amwâs. The main range trends to the south-west and west, and has on its summit the villages of Kuriet el Enab, Saris, and Beit Mahsir. It is probably the Mount Ephron of Joshua xv, 9; Kuriet el Enab being Baalah or Kirjath Jearim. A road from Ramleh by el Kubab, and between Latron and 'Amwâs, passes up to the summit of Saris, and follows the ridge to Kuriet el Enab, where it turns eastward, passing Kustul, Kulonieh, and Lifta, to Jerusalem.

(2.) The orography of the middle part of the group will now be described. It is all within the upper portion of the Nahr Rubin Basin, and its watercourses are noticed under that heading.* The northern extremity of this tract, and indeed the whole of its eastern side, is included in the western plateau of Jerusalem, with the Plain of el Jib on the north, and the Plain of Rephaim or the Giants on the south. It will be remembered that the Wadys Ahmed, Sikkeh, and Surar or Ismain, form in succession the southern limit of the tract. The spurs or ranges thrown off from the 'Auja-Rubin waterparting towards the Surar, will be noticed first. Those from the Jerusalem Plateau, to the left bank of the Surar, will follow. It will be convenient here to apply the name of Wady el Surar to the central main wady from Lifta downwards.

The connection of the conical mountain of Neby Samwil with the main range has been already noticed. The proper base of this mass is well defined by the Wadys Amir, Beit Hannina and Buwai. On the right bank of the Buwai, a distinct range proceeds from the waterparting on the west of Biddu, and descends to the Surar at Kulonieh. The same range is prolonged westward to the junction of Wady Ghurab and el Mutluk, with Wady es Surar, and separates the

* See *ante*, pages 42 to 45.

deep and rugged gorges of those wadys. It is believed to be the Mount Seir of Joshua xv, 10. On its summit are found the villages of Beit Surîk (alt. 2,690 feet), Kustul (alt. 2,650 feet), Soba (alt. 2,567 feet), Kh. Bâtn es Saghir* (alt. 2,280 feet), Kh. Shufa (alt. 2,697 feet), 'Akur, and at its extremity is 'Artûf (alt. 910 feet). Below 'Artûf, the valleys expand into a beautiful plain, surrounded by many biblical sites.

Spurs of Mount Seir or Saghir descend to Wady Ghurab on the north, and to Wady es Surar on the south. The most important is one that divides Wady esh Shemmârîn from Wady es Surar. It contains the villages of Setaf and Kh. el Loz. The altitude at the confluence is 1,297 feet. Another spur descends from Akur to the left bank of Wady el Hamar or Ghurab, and it includes the village of Nesla, the Chesalon of Joshua xv, 10.

Other spurs descend from the main range between Beit Surîk and Kuryet el Enab; but they are only short corrugations of the right bank of the Upper Ghurab. Lower down, that wady hugs the main range closely. It is reported to be a deep and narrow chasm, though less deep and wild than the parallel Wady Ismain or Surar, on the south.

There are no spurs emanating from the western side of the Jordan waterparting, or from the Jerusalem Plateau on the westward of it, to the northward of Lifta; for the swelling ground about Shafat is scarcely of that description. South of Lifta, the course of the Wady Surar is advanced suddenly westward for about two miles, and thus space is found for the commencement of tributaries to Wady el Werd and es Sikkeh, as well as of the ridge which separates that wady throughout from the Surar. It is called Jebel Ali. Along its slopes towards Wady Surar, are the villages of Deir Yesin, and Ain Karim, the latter having the Latin monastery of St. John.

(3.) The Jordan watershed of the present group extends

* This name is the Arabic equivalent of the Hebrew Seir, both meaning rough, rugged.

from Bethel to Bethlehem, and from Wady Nûei'ameh on the north of the Plain of Jericho to the Wady Derajeh on the Dead Sea. Between Nûei'ameh and Derajeh are the basins of Wady Kelt, Wady el Kueiserah, and Wady en Nar or Brook Kidron: besides three groups of secondary basins.*

The plains which distinguish the western and central parts of this group, are not without a representative among the rugged features of the eastern side, where the Bukei'a extends between Wady Mukelik and Wady en Nar, about two miles from the Convent of Mar Saba, and above the line of cliffs at the northern end of the Dead Sea.†

The bottom of the mountains between 'Ain Nûei'ameh and Wady Kelt is a wall of rocky precipices, which rises at Jebel Kuruntul or Quarantania to more than 1,000 feet above the Plain of Jericho, or 320 feet above the sea. Between Wady Kelt and El Kueiserah, a distance of four miles and a half, the cliffs give way to a slope which rises about 2,000 feet above the plain, to Talat et Dumm on the road between Jericho and Jerusalem, and about four miles from the plain. The precipitous wall is the abrupt termination of ridges that descend rapidly from the line of Eastern Summits that was brought down in the preceding group to Kubbet Rummâmaneh (alt. 2,024 feet). From that point the Eastern Summits are carried on by a distinct line of heights running southward to Ras et Tawil (alt. 1,964 feet); and from thence by another range at the head of Wady Rijan, to a summit on the south of the confluence of Wady es Suweinit, with Wady Farah. This summit is on the waterparting of Wady Kelt, which the line now follows to Talât et Dumm. From Talât et Dumm, the summit of this slope runs for two miles southeastward to Jebel Ekteif, which is only three miles from the plain, and about 1,800 feet above it, or 840 feet above the sea. The slope is included in the secondary basins on the north of el Kueiserah, described before at page 95. South of el Kueiserah, the cliffs reappear, and skirting the Dead

* See pages 84 to 106.
† See page 99.

Sea, they form the base of the highland for the rest of the survey. They rise to a height of about 1,000 feet, and between el Kueiserah and Wady ed Derajeh, they are capped by summits that slope upwards to a further altitude. Tubk el Kuneiterah is 306 feet above the ocean, or 1,598 feet above the Dead Sea. Tubk es Sammarah is 530 feet above the ocean, or 1,822 feet above the Dead Sea. These heights are on the eastern side of the Plain of el Bukei'a, which has on the west a range which culminates in el Muntar, a mountain on the north of the Greek convent of Mar Saba, with an altitude of 1,723 feet above the ocean, or 3,015 feet above the Dead Sea. The range of el Muntar is a prolongation southward from Jebel Ektief, and it is intersected by the rocky gorge of Wady Mukelik at the southern foot of Ekteif. From el Muntar the range is continued to the south-west, and crosses the Wady en Nar where it bends to the south, and soon runs due south by Akabet el Murajeh (alt. 1,600 feet) to the Wady el 'Alya, and passes it by Muksar Ismain in a south-westerly direction to Wady Tamireh, beyond which it follows another ridge running due south to Wady el Muallak or ed Derajeh, the southern boundary of this group. This line of Eastern Summits will be found extending to the end of the survey, oftentimes and generally with a distinctness which throws a light on less obvious portions.

On the north of the Plain of el Bukei'a, the slopes between the Eastern Summits and the cliffs, is a succession of ridges and ravines with the rocky gorges of Wady Rummâmaneh, and Wady Kelt on the north and south. A highway from Jericho, through Mukhmas and Deir Diwan to Bethel crosses this part. The highway from Jerusalem to Jericho enters the slope at Talât et Dumm, which is therefore a noted point; and it is considered to be identical with the Adummim of Joshua xv, 7. Between el Bukei'a and Wady ed Derajeh, the slope passes from the rugged valley of Wady en Nar or Kidron, to rolling chalk downs spreading out from the lofty cliffs that border the Dead Sea to the Eastern Range; which here throws out upon the downs, like advanced posts, the heights of Kurn el Hajr (alt. 1,460 feet), and er Rueikbeh (alt. 1,486 feet). One of the roads from Jerusalem to Engedi, crosses this part near the top of the cliffs.

The tract between the Main Waterparting Range and the Eastern Range, as far as the summit on the south of the confluence of the Farah and Suweinit, may be distinguished as the northern part of the Eastern Plateau of Jerusalem. South of the unnamed summits, the eastern edge of the plateau must be sought further to the west than the Eastern Range, as it has been traced on the south of that summit. The westerly projection of the Jordan Waterparting, as it advances southward, widens the slope; and the altitude and structure of the ground together indicate the interposition of a medial range and terrace, which may be observed all the way southward from Wady Farah.

The Middle Range passes from the summit on the south of Wady Farah, along the waterparting south-westward, to the rocks of 'Arâk Ibrahîm; when it proceeds southward, across two wadys, to a long range distinguished by Arak esh Shem, Kh. Karrît, and Kh. er Raghabneh. The Middle Range is now intersected by a remarkable bend of Wady Abu Hindi, and proceeds to a similar bend of Wady Abu Nar, both of the bends being due to the intersection of the range and the sudden descent of the wadys from higher to lower ground. The range is next defined by the prominent heights of Deir Ibn Obeid* and Umm el Tala, on the east of Bethlehem, and crosses the Wady el War to Jebel Fureidis, or Paradise, called also the Frank mountain, and the site of the fortress of Herodium erected by Herod the Great, on the southern boundary of the present group.

The Middle Range from Wady Farah to Jebel Fureidis, and the Eastern Range from Wady Farah northward to Ras et Tawil and Wady Rummâmaneh, separate the habitable tract of the Eastern Plateau of Jerusalem, from the wild ravines and ridges that descend from it rapidly and form the lower parts of the mountain. The plateau is generally

* The ruined Monastery of Theodosius, including two churches.

more than 2,000 feet above the sea, up to its eastern edge in the Middle Range. It contains Jerusalem and Bethlehem, with several villages and cultivated grounds, besides many ruined sites that are the remnants of a much larger population. Between the Middle and Eastern Ranges, and below the latter to the foot of the mountains, not a village is to be found, and the only settled habitation is the fortified Greek Convent of Mar Saba. The high lateral valleys on the east of the main waterparting that have been traced from the north of Samaria up to Wady el Ain and the northern boundary of this group, are continued southward from Deir Diwân, to Jeba, Hizmeh, Anata, and el 'Aisawîyeh to Jerusalem. But the lateral character of the valleys on the Eastern Plateau is much less developed, than it is in the valleys of the Western Plateau, from Bireh to Bethlehem. South of Bethlehem, the lateral valleys are found again chiefly on the eastern side of the range as far as Hulhul; when they are once more transferred to the other side, and are taken up by the great Wady el Khulil and its parallel affluents.

The Hill Country of Judah, or the Hebron Group.

The eastern boundary of this mountain system is defined by the shores of the Dead Sea. On the south the subject cannot for the present be discussed owing to the limits of the Survey. On the west the boundary has been already indicated southward from Wady Surar along the Wady en Najil and Wady es Sur. South of the head of the latter wady, and as far as Beit Auwa, there might be some doubt about the continuity of the meridional division between the foot of the Mountains of Judah, and the Lowland Hills of the Shephelah, as the hill drawing of the Survey is not so expressive or accentuated as it might be. But here the evidence of Dr. Robinson removes all hesitation as far south as Tell Beit Mirsim and Burj el Beiyarah.* Indeed the

* He remarks that "Idhna lies at the foot of the mountains, where the steep ascent of the higher ridge soon commences." "Bib. Res." ii, 70.

meridional valleys occur again at Beit Auwa, where they spread out into a plain stretching southward as far as Tell Beit Mirsim, and are continued by Wady el Beiyârah and Wady edh Dhikah up to Khurbet Khuweilfeh. Here is the waterparting between the basins of Wady el Hesy and Wady Ghuzzeh, and the boundary of the Shephelah now turns abruptly to the west, along the line of Wady Khuweilfeh and the great Wady esh Sheriah, to Wady Ghuzzeh and the sea. South of Khurbet Khuweilfeh, the Hill Country of Judah drops down to the broad Plain of Beersheba. For the sake of precision, the dividing line between the Judean Hills and the mountain that divides the Plain of Beersheba from Wady esh Sheriah, may be traced along a wady descending from Khurbet Khuweilfeh, by Wady Itmy to Wady el Khulil, and its outlet into Wady es Seba.

The summit of the Main Range of this group coincides with the Mediterranean and Jordan waterparting from Bethlehem southward. It forms the eastern boundary of the basins of Nahr Rubin, Nahr Sukereir, and Wady el Hesy, and has been already traced in connection with them.*

The Western Slope.

The northern side of the western slope begins its descent to the northern boundary of the group formed by Wady Ahmed, Wady el Werd, Wady Sikkeh, and Wady es Surar, from the main waterparting between Bethlehem and el Khudr, where it is spread out between Wady Ahmed and Wady Bittir. From el Khudr, a range strikes westward by Hausân, and Deir Hawa, to terminate at the confluence of Wady Ismain or Surar with Wady en Najil. It is followed by a highway along the eastern part of the ridge, and by a track along the western part. It sends only a short slope to the wadys of the northern boundary; but it throws off considerable spurs to the south. Thus from Hausân a short spur forms the upper part of the left bank of Wady Musurr

Also, "The ruins of el Burj are situated very near the border of the hilly region towards the western plain." "Bib. Res." ii, 218.

* See *ante*, pages 46, 51, 54.

and a longer one reaches to the lower part where it joins Wady el Jindy. The next spur divides Wady el Jindy from Wady el Werd, and terminates at the head of Wady en Najil. Five more spurs descending from the ridge to Wady en Najil complete its southern slope.

Between el Khudr and Balutet el Yerzeh, the main range descends to Wady Musurr, and from the Balutet a long range extends westward to Wady es Sunt, and divides Wady el Jindy from Wady Helwas. It has the small village of Jeba on its summit. Between Balutet el Yerzeh and Khurbet Jedûr, the main range descends by short spurs to Wady el 'Abhar, a branch of Wady Helwas. The villages of Safa and Surif are on this part of the slope. Between Khurbet Jedûr and Hulhul, all the ridges descend to Wady es Sur or its affluents; and the villages of Kharas, Nuba, and Beit Aula, are situated in a line on the lower part of the slope. A ridge running direct west from near Hulhul to the head of Wady Bir es Suweideh, and then bending northward along the left bank of the latter wady, divides the affluents of Wady Sunt from those of Wady Afranj, both being in the Sukereir basin. From near Hulhul to Dura on the main range, the spurs fall to Wady el Afranj. At Dura, the advance of the main range towards the west, reduces the length of the western slope; and the contraction gradually comes to a point at Kh. Khuweilfeh from the same cause. A north-westerly spur from Dura divides the Afranj from the affluents of Wady el Ghueit, which is the southernmost of the great divisions of the Nahr Sukereir basin. Only a mile south of Dura, the waterparting between the Sukereir and el Hesy basins emanates from the main range at Ras el Biain, and is formed by a long range running westward to the sea at Ascalon. It is the slope drained by the affluents of the Ghueit, between Idhna and Beit 'Auwa, which breaks the continuity of the meridional valleys, that form the division between this mountain group and the Shephelah. But a careful study of the Survey, coupled with Dr. Robinson's notes, leads to the conclusion that in this part the separation will be found as distinctly marked as elsewhere, and perhaps more broadly.

Between Ras el Biain and Shaik Abu Kharrubeh, all the spurs descending from the main range between those points, which are about eight miles apart, concentrate in the plain on the east of Tell el Akra. Those also between Kharrubeh and Kh. Khuweilfeh meet in the plain called Sahel es Sabti. The descent from this point to Beersheba belongs to the other side of the main range. The bold and bluff descent of the western hills to the valley which divides them from the foot of the eastern range, is well shown on the new maps in this part.

The Eastern Slope.

The northern side of the eastern part of the present group, the Hill Country of Judah, descends to Wady Khureitun from a range which emanates from the main waterparting at Ballutet el Yerzeh (alt. 3,167 feet) on the east of the village of Safa. The range first proceeds for a mile and a half to the south-east, and then turns a little east of north for a like distance, when it bends eastward to reach the high plateau of Tekua. The bend at starting is made to turn the head of Wady el Biar, which descends the northern slope for five miles in a north-easterly direction, and joins the boundary of the group in Wady et Tahuneh, on the south of Bethlehem. From the plateau of Teku'a, the northern range extends to the south-east, and ends in Ras Nukb Hamar which drops by lofty cliffs to the south side of the gorge of the Derajeh where it opens on to the shore of the Dead Sea. The Ras is 678 feet above the ocean, or 1,970 feet above the Dead Sea.

A ridge on the right bank of Wady el Biar, runs north-eastward to Wady et Tahuneh, and then bends to the south-east along that wady, as far as its junction with Wady Abu Nejeim, which rises near the head of Wady el Biar. Wady Abu Nejeim drains the southern side of this ridge, and also the northern side of the Plateau of Teku'a. On the eastern side of Teku'a are the great Caverns of Khureitun, in a cliff overlooking the Wady Khureitun, which is a part of the northern boundary of the present group. The caverns face Jebel Fureidis; and in the time of the Crusaders they came to be regarded as the Cave of Adullam; but the true site lies far away at the western foot of the mountains, at Kh. 'Aid el Ma, in the Wady es Sur. Teku'a is on the waterparting between the basins of Wady ed Derajeh and Wady el Areijeh, the latter emptying itself at Engedi. From this waterparting near Teku'a, three wadys with intermediate spurs descend, and unite in a rocky gorge which cuts through the northern ridge and joins the gorge of ed Derajeh.

The plateau of Teku'a is the first part of the Middle Range in this group. Its prolongation to the south is found in a very distinct form in the long ridge of Kanân ez Zaferan, which, with the plateau of Teku'a, encloses the heads of the basin of Wady el Areijeh, that have their sources along the main range from Safa to Hulhul. These head valleys all concentrate in the Wady el 'Arrûb, which soon after receives the drainage of a circular tract on the west of the plateau of Teku'a, and then cuts through the Middle Range on a rocky gorge. The breadth between the Main and the Middle Ranges at the widest part, where the 'Arrûb enters the gorge, is about six miles. Near Hulhul it contracts to two miles, but it widens out again immediately to three miles, and goes on with a breadth of about four miles.

The Kanân ez Zafaran terminates at Khurbet el Addeiseh, on the waterparting that divides the basins of el Areijeh and Ghuzzeh, at the head of the valley of el Khulil or Hebron. The Middle Range passes from the Kanân to this waterparting range, and follows it to Beni Naîm (alt. 3,120 feet) and on to Khurbet Birein. Between Beni Naîm and Khurbet Birein the waterparting divides the affluents of Wady el Khulil from those of Wady el Khubera. At Khurbet Birein, the Middle Range has its summit on a ridge that contains the ancient town of Yutta or Juttah, and divides from the affluents of Wady el Khulil, the wady called

R

at its head Wady Abu Hirsh and Wady el Butm on its lower course. The Middle Range comes to an end in the Plain of Beersheba.

From the head of the Hebron Valley southward, the space between the Main and the Middle Ranges, is wholly occupied by that valley and its branches, which throughout maintain an average width of four miles. The length of the valley from its head to its junction with Wady es Seba, is about 30 miles.

Eastward of the Middle Range, and parallel to it, is the continuation of the Eastern Range, from the northern boundary of the group at Wady el Muallak. The line passes from Kubr Ghaunameh, to the head of the rocky gorge of Wady Mukta el Juss, thence by the spurs that separate the Wadys Dannûn and Bassas to the range at el Megheidhah, and Rujm Abu Zumeitir; thence to the summit of Dharet el Meshrefeh (alt. 1,696 feet), and southward to Dharet es Sukiyeh (alt. 1,836 feet), Dharet el Kolah, and the summit of the range on the south of Wady el War, which runs unbroken for 23 miles to the Wady el Milh. The principal summits on this very distinctly defined and concluding portion of the Eastern Range, are: Tell et Tuany (alt. 2,837 feet), on the highway between Yutta, Kurmul, and Engedi; and Kanân el Aseif (alt. 3,002 feet), on the road from Yutta to Tell el Milh, or Moladah.

This first attempt to indicate, and to give a precise limitation to the triple succession of terraces above the Dead Sea cliffs, which the New Survey has disclosed, may not be always supported by a critical examination on the ground in every particular; but generally the distinct separation of the three steps, seems to be unquestionable. The lowest step between the cliffs and the base of the Eastern Range is about four miles wide at the north end, and gradually increases to about eight miles in the southern part. Only a few scattered summits rise to 1,400 feet above the ocean, but it must be remembered that the surface of the Dead Sea is 1,292 feet below that level. Between Teku'a and Ain Jidy, in this step, is probably the Wilderness of Jeruel, 2 Chron. xx, 16. "The end of the brook," may be the head of Wady el Mukeiberah, the invaders being drawn up on the slope facing the ascent to Teku'a.*

The steep ascent to the next higher level forms part of the Eastern Range, the length of which, south of Wady ed Derajeh, exceeds 30 miles, and for more than two-thirds of that length, the range has a descent on the western side, as well as the eastern. In other parts, it is but an acceleration of the slope to a greater or less extent, but it seems to be always sufficient to maintain the continuity of the feature.

The southern part of the middle terrace, or the step between the Eastern and the Middle Ranges, includes several villages at the present time, and some ruins, among both of which are representatives of the very ancient towns of the Hill Country of Judah; such are Kh. Attir or Jattir, Semua or Eshtemoh, Kurmel or Carmel, Tell Main or Maon, Yuttah or Juttah, Tell es Zif or Ziph, Kh. Gannim or Anim, Beni Naim or Janum, the latter names, in each case, being biblical.

North of Ziph, the habitable country recedes to the Middle Range, or westward of it; for on the east the country is too steep or broken, and only fit for nomadic pasturage. The triangular tracts, enclosed by ranges of hills on the east and west of the Kanân ez Zaferan, on the north of Hebron, are curious contrasts; the western being a populated plateau, while the eastern is a remarkably steep slope and probably uninhabitable, and only fit for pasture.

The highest plateau which lies between the middle and main ranges is habitable throughout, although the existing villages are sparse and small; but the evidences of a much larger population in former times, are everywhere, both here and on the western slopes. Its principal features, beginning on the north, are (1) the long Wady el Biar, which skirts the Main Range from Kh. Breikut, to Urtas on the northern boundary of the group. (2) The widely extended branches of Wady 'Arrûb, with which may be noticed the more northerly

* See *ante*, page 109.

valleys uniting in Wady Rekeban, and those further north that join in Wady Abu Nejeim. The spurs of these valleys are threaded by an aqueduct, which draws its supplies from Birket Kufin, near Beit Ummar, and from 'Ain Kueiziba, about two miles on the south-east; these unite at Birket el 'Arrûb, and the aqueduct then passes along the left bank of Wady 'Arrûb, till it approaches the confluence of Wady er Rekeban, when it turns northwards, and meanders up and down the sides of valley after valley, till it reaches el Burak, near Urtas. (3) The valley of Hebron or Wady el Khulil. The northernmost head of this great valley commences, as Wady en Nusara at Khurbet Beit Anun, the Beth Anoth of Judah, near Hulhul.

The Wady el Khulil originates (1) partly on the eastern and southern slopes of a rectangular bend of the waterparting that divides the wady from the basin of Nahr Sukereir, on the west, and (2) partly along the Mediterranean waterparting, where it makes a rectangular bend eastward, and then to the south-east, between Hulhul and Beni Naim, dividing the Khulil from the Areijeh Basin.

The eastern and southern slopes respectively, of the first-named rectangular bend, have their drainage divided by ground thrown off as a ridge from the projecting corner of the bend, in a south-easterly direction, and then expanding like a fan, which terminates in short slopes on the north and west, and in a long slope towards the south-east and south. This divides the head of the valley into two parts, and the eastern and larger part slopes towards the waterparting that runs from Hulhul and Beni Naim, close along which the channel passes which collects the waters until spurs from Beni Naim, and further south, push the channel against the long slope of the fan before-mentioned, at the southern end of which the eastern channel unites with Wady Khulil in meandering southwards. The western part is limited on the south by a ridge extending eastward to the foot of Hebron, and throwing off a long branching spur to the south, as far as Rujm ed Deir. Below Hebron the ridge is divided by a gorge from the fan before-mentioned. There is, at the head of the eastern part, a valley in the centre, and others along the eastern and western sides of this part, all running southwards, till the western is diverted towards the head of the gorge by the southern ridge, and receives the others on the way. The gorge takes the name of Wady el Kady, or el Khulil, and after a meandering course southward for more than a mile it bends abruptly eastward, and meets the eastern branch. After the junction, the main wady again meanders to the south, till it is turned to the south-west by the Yutta range, and the spurs proceeding from it, which supply two considerable tributaries to the left bank, the first at Khurbet Rabut and the second at the foot of a spur surmounted by a road from Dhaheriyeh. Here the Wady el Khulil again impinges on the descending Yutta Range, and winds around its short bluffs, down to the Plain of Beersheba.

On the western side, from the waterparting at Khurbet Kanân, an inner range is given off, which forms a long plateau or narrow terrace with lateral valleys between it and the waterparting. It encloses the branching valleys at the head of Wady ed Dilbeh, and forms the unbroken range, about six miles long, on which Dhaheriyeh is situated. For two miles and a half in the central part the continuity of the range is only traceable in more rapid slopes without lateral valleys; but this does not prevent the highway from Hebron through Dhaheriyeh to the Plain of Beersheba, from following this inner range throughout.

The Shephelah or Philistia.

The natural limits of the Shephelah have been discussed in the foregoing chapter. Viewed as a whole, these hills present the aspect of an amphitheatre, encircling the Plain of Philistia on the east, from Gaza to Jaffa. The new survey which has enabled the eastern boundary to be defined, also supplies the means of unriddling the tangled structure of the mass, and to that subject the following remarks will be addressed.

A succession of main wadys from the mountains of Judah, intersect the hills of the Shephelah from east to west, and constitute dividing lines between the five groups into which it is convenient to arrange the mass for the purpose of its description.

I.

The Wady es Surar is the most northerly of these dividing wadys. From its northern or right bank, a range of hills extends in a north-westerly direction nearly up to Jaffa. The villages of Eshua, Amwas, el Kubab, the towns of Ramleh and Ludd, with Safiriyeh, Beit Dejan and Yazur, skirt the northern foot of the range towards the Judæan Hills and the Plain of Sharon. On the summit of the range are Surah and Abu Shusheh. On the western side are Kuldeh, el Mansûrah, Naaneh, Akir, el Mughâr, Zornukah, el Kubeibeh and Surafend. The waterparting between the basins of Nahr el 'Auja and Nahr Rubin, is on the summit of this range.

The range is distinctly divisible into three parts, of which the southernmost has a biblical celebrity derived from the birth and exploits of Samson. It is a circular block surrounded by Wady Surar, Wady el Khalil, and Wady Atallah, the last flowing by Latrôn. From a semicircular outer range, rising from the south-eastern part of the base two main valleys pass through the block westward, and fall into Wady es Surar. It has been already noticed on p. 50 that the southern valley contains the ruins named Khurbet Surîk, and that name combined with the situation of the valley, serves to identify it with the Valley of Sorek where Samson was taken prisoner in the house of Delilah. Zorah, where Samson was born, is the present village of Surah (alt. 1,171 feet) at the head of the valley and on the summit of the outer range. The northern valley is of greater extent and drains the main body of the block. The site of Samson's En Hak-kore is the Ayun el Khaijeh, and Ramath Lehi is placed at Kh. Ism Allah. Perhaps the site of the Mahaneh or camp of Dan is Khurbet Kila, or the ruined fort. Spurs from the outer range divide it from the Vale of Sorek on the south and from the Wady el Khalil on the north. The greatest length of this block is about six miles, and its breadth exceeds four miles.

On the north of Wady el Khalil is another division of the range, with the Wady Harir on its north-eastern base, and the Plain of Akir or Ekron on the south-west. The north-western boundary is traced from the Plain of Ekron along Wady Bahlas, to an affluent of Wady Harir, having Ras Abu Hamid at its outlet. The only biblical site on this block is at Tell Jezar identified with Gezer by M. Ganneau, and having now in its neighbourhood the modern village of Abu Shusheh. The natural character of this mass is in contrast with the first, for the highest part instead of being on the exterior, is in the centre where it forms a crater-like summit from which the slopes descend on all sides. The highest point is 850 feet above the sea.

The northernmost and third division of this series consists of a low range which stretches from el Mughar northward to Beit Dejan, a distance of twelve miles, and spreads westward to the sandy downs which separate the hills from the sea. This range forms the western boundary of the Plain of Ramleh and Ludd on the north and the Plain of Ekron on the south. One of its central valleys is Wady Deiran, a word probably derived from Daroma, an ancient name of this district. The highest part of this undulating track is 261 feet.

II.

The next division includes the hills between Wady es Surar and Wady es Sunt, which may be reckoned on an average about four miles apart. Here also are some distinct features. On the east there is a ridge extending from the Sunt to the Surar, with a short slope of about half a mile towards the eastern limit of the Shephelah in Wady en Najil; and a longer slope of about two miles towards the west, where the drainage is received by a valley, generally parallel to Wady en Najil, and also running northward to Wady es Surar. At the head of the valley is Beit Nettif (alt. 1,517 feet) with a fine view; in the centre is Beit el Jemâl; and at its outfall is 'Ain Shems (alt. 917 feet), the Beth Shemesh of Scripture (1 Sam. vi, 9–20).

The heights on the west of the valley also extend from the Sunt to the Surar. But these heights throw off spurs and ranges to the westward. The southern part descends in short spurs to Wady es Sunt where the wady makes two great bends to the north.

Then a long range is thrown off to the west, beginning on the south of Khurbet el Kheisham (alt. 1,245 feet), and having on its summit the village of Mughullis and Kh. el Mensiyeh, where the range bends round to the north, having the village of Dhennebbeh on its western flank, which descends to the plain. On the north of Dhennebbeh, the range is indeed prolonged westward across the plain at a low elevation, but sufficient to make it the continuation of the waterparting between the basins of Nahr Rubin and Nahr Sukereir, which entered this division at Beit Nettif. From its origin to Kh. el Menshiyeh the range skirts the right bank of Wady es Sunt and divides it from Wady el Menâkh, at the head of which is the village of el Bureij (alt. 830 feet).

North of Khurbet el Kheisham, another range is extended to the west, between Wady el Menakh and Wady es Surar. It terminates on the plain where the wady enters it.

III.

The hills between Wady es Sunt and Wady el Afranj, now come under notice. They arise on the east in a range which originates between the head of Wady es Sur and Wady el Afranj. The range forms the left bank of Wady es Sur, which is a part of the eastern boundary line of the Shephelah, following it northward till the Sur joins the Wady Musurr, and from the union proceeds Wady es Sunt. The range now turns to the westward as far as the junction of Wady Zakariya with the Sunt.

The descent of the range eastward to the Sur is abrupt, like that of the range along Wady en Najil on the north. The slope to the west is much longer, and about four miles in width, terminating partly in Wady el Afranj on the north-east of Beit Jibrin, but chiefly in Wady Zakariya, called also Wady es Seiji, which rises about two miles north-eastward of Beit Jibrin, and runs northward to Wady es Sunt.

At the foot of Beit Jibrin, but on the north of Wady el Afranj, a range commences, which runs northward to Wady es Sunt, where it terminates in Tell Zukariya. This range is altogether seven miles in length, and completely separates the eastern tract, just described, from the western. The range rises steeply on the east, from the banks of Wady el Judeiyideh, and Wady Zakariya; but it throws off long slopes towards the plain, which however they are prevented from reaching by the interposition of another range running from the right bank of Wady Afranj, on the north-west of Beit Jibrin up to Wady es Sunt, where it ends in Tell es Safi. This western range throws the southernmost waters of the middle range to the north, and collecting the rest of the western drainage of the middle range in them, finds an outlet to the Sunt in Wady es Safi, on the east of Tell es Safi. The middle range rises to 1,335 feet in Kh. es Surah. On its western slope are the villages of Kudna and Rana, and Deir edh Dhibban.

Besides Tell es Safi (alt. 695 feet), the summits of the western range are distinguished by the village of Dhikerin (alt. 680 feet), and its slopes contain the villages of Berkusieh (alt. 585 feet), and Summeil (alt. 405 feet). From the ground between those villages a slight swell extends across the plain to the villages of el Butani, about five miles from the sea.

IV.

This group has the Wady el Afranj on the north, and the main channel of Wady el Hesy on the south. This main channel enters the Shephelah at the junction of the Wady el Butm, through the gorge of Tell el Akr'a; and it emerges at Tell el Hesy into an arm of the western plain, having the villages of Bureir and Simsim at its mouth.

The village of Idhna is the central point of the northern part of these hills. From thence a range runs north-westward, following the curvature of Wady el Afranj, and forming its left bank, until it reaches Khurbet Senâbreh, when the range is deflected to the south-west, along the northernmost affluent of the southern branch of the Sukereir Basin, which crosses the plain as Wady el Ghueit. The range now divides the affluents of Wady el Afranj from those that fall to el Ghueit, and in the performance of this function it reaches the plain on the south of Zeita. The most noted place on its summit is Tell Sandahannah on the south of Beit Jibrin.

Another range proceeds from Idhna westward to el Khubeibeh, and throws off considerable spurs to the northwest, which terminate on the left bank of the Ghueit affluent before mentioned, called in part Wady Beit 'Alam, and filling up the space between the two Idhna ranges.

The next range arises between the wadys that rise about Beit 'Aûwa. It is in continuation of the waterparting between the basins of Nahr Sukereir and Wady el Hesy, which begins at Ras Biain on the south of Dura. It throws out spurs on the south side of the range, from a point on the south of ed Dawâimeh. They are divided by the Wady el Butm; and at its western end it sends two spurs towards Bureir. Otherwise it hugs the right bank of Wady el Hesy so closely, that no features of that kind occur. To the basin of Nahr Sukereir, this range contributes not only spurs, but a considerable branch, which, beginning on the west of ed Dawaimeh, passes to the plain at Arak el Menshiyeh. From the main range further west, a spur passes between Wady el Habur and Wady es Sukriyeh; and between the latter and Wady en Neda, a broad slope descends to Khurbet Fattatah. Tell Ibdis caps the termination of the main range, which falls boldly to the plain northward from that summit between Arak el Menshiyeh and Khurbet el Jils.

V.

South of Wady el Hesy, the next dividing line is found in Wady esh Sheriah, and its upper course called Wady Khuweilfeh. Along the eastern margin of this extremity of the Shephelah, a range rises from the banks of Wady el Butm and Wady el Beiyarah, which extend between the plains of Aitun and Sabti. From the central points of the range at Kh. Jeimer (alt. 1,530 feet), a range proceeds north-westerly to Kh. Surrar, presenting an abrupt escarpment to form the left bank of Wady el Jizair or el Abd, and a long slope with tributary streams towards the left bank of Wady en Nas.

South of the Plain or Sahel es Sabti, a long range emanates from the waterparting between Wady el Hesy and Wady Ghuzzeh at Kh. Bureideh. On its summit on the south of Sabti is Kh. Z'ak (alt. 1,370 feet), followed by Jebel Abu Huteirish, and the Ard el Mak-huz, the termination being at Tell el Hesy. The only notable spurs on the north-east side of the range are near the Ard el Mak-huz. On the south-west side they occupy a considerable slope which descends to the Wady el Muleihah. The country is entirely pastoral with some ruins, but no existing villages.

The next range is part of the waterparting between the Wady el Hesy and Wady Ghuzzeh, beginning at Kh. Khuweilfeh and proceeding westward with bold curves to the maritime plain on the south of Gaza. It has been already traced.* Its northern side is closely followed for about six miles by the Wady Nuksar; indeed so closely that the wady may almost be said to be on the summit. After that wady turns to the north-west, five spurs descend in succession to Wady el

* See *ante*, pages 54, 55.

Muleihah and Wady el Hesy. The most westerly of these spurs, throws off several very rugged arms westward to the plain. From Sateh Burber on the main range, a considerable branch extends to the villages of Huj and Nejed in the plain, spreading out its lower features towards Gaza. North of Nejed a group of low hills of a similar character divide the plain between Simsim and Keratiya from the margin of the sea, where the famous city of Ascalon once held its powerful sway. Westward of Sateh Burber, the main range nowhere reaches an altitude of 500 feet, and the undulations do not invite much notice. The same may be said of the slope from the main range towards the Wady esh Sheriah. Judging from the map it exhibits the ordinary aspects of chalk downs. It appears to be passable in every direction, if the numerous tracks over it may be taken as evidence to that effect. The hills to the south of Wady esh Sheriah within the Survey, appear to be of the same character, and seem to be deprived of interest, owing to the obscurity in which the south country or Negeb is still left. That the country will ultimately be surveyed to which Abraham bent his steps in fulfilling to the utmost the divine command,—the Gerar which became his home, and the scene of his exemplary confidence in the presence of the Almighty,—the Negeb with its interesting events in the life of David,—and the southern border of the Promised Land with the long disputed site of Kadesh, and many other places the sites of which Mr. Wilton has so ably discussed,—that this portion of the Land of Promise, may be added to the Survey is ardently desired.

www.ingramcontent.com/pod-product-compliance
Lightning Source LLC
Chambersburg PA
CBHW032006230426
43672CB00010B/2269